ENCYCLOPAEDIA OF ENTOMOLOGY

Encyclopaedia of Entomology

Volume 2

Insect Control

Bharat Singh

ANMOL PUBLICATIONS PVT. LTD.
NEW DELHI - 110 002 (INDIA)

ANMOL PUBLICATIONS PVT. LTD.

H.O.: 4374/4B, Ansari Road, Daryaganj
New Delhi - 110 002
Ph.: 23261597, 23278000

B.O.: No. 1015, Ist Main Road, BSK III Stage
III Phase, III Block
Bangalore - 560 085 (India)
Visit us at: www.anmolpublications.com

Encyclopaedia of Entomology

First Published, 2006

ISBN 81-261-2610-8 (Set)

PRINTED IN INDIA

Published by J.L. Kumar for Anmol Publications Pvt. Ltd., New Delhi - 110 002 and Printed at Mehra Offset Press, Delhi.

Contents

Preface

Science is the sermon of today. In every walk of life, since childhood till old age, science plays an important role in the life of man. Uptill some decades back, studied merely as a subject in schools and colleges, Science has assumed the status of an essential part of life today. With the expansion of literacy among the masses and growing awareness, they have woken up from the slumber of ignorance and have started taking up things more sincerely and thoughtfully. It is one of the reasons that our nation is heading towards new goals day after day.

It is a vast subject to study and talk about. For the purpose of ease, Science has been divided into more than two hundred and fifty branches. Specialisation in any branch can be obtained at different institutions and universities, scattered all over the country and abroad.

Apart from other super-specific branches, Physics, Chemistry, Zoology, Botany, Mathematics and Computronics, etc. are more general and closer to daily human life. From the point of view of the cut-throat competition in educational world, a thorough study of the above-mentioned branches, proves very handy, for the new aspirants, especially and for teachers and tutors, generally.

An urgent need was being felt for such a series of books, encompassing the commonly-studied and discussed branches of Science of equal significance and presented in an easy language, understood by all. With God's grace, our dream came true and we are happy, while introducing the dream-series for the lovers of knowledge.

The series consists of many a title, covering various branches of Science, such as Physics, Chemistry, Zoology, Botany, Geology, Agriculture Science, Computer Science, Mathematics and so on. The series is important and a must not only for students but also for teachers. Explained in simple language and aided with graphics and diagrams, everything, between the two ends—title and back cover—of the series, has been presented in a perfect way.

This set of books is devoted to **Entomology**, which comprises two volumes: (1) Insects and Agriculture and (2) Insect Control.

It is hoped that the series would serve both, the students community and the academics, equally and will prove to be a boon for those, aspiring for newer heights. Nevertheless, any suggestions for further betterment of the endeavour will be highly appreciated and solicited.

Editor

1

Pest Control

Insect control in broader sense may be defined as to make the life of an insect harder or to prevent their increase or spread or to kill them by any means. It depends upon the knowledge of the factors causing or favouring the abundance of the particular species. The control of insects can be achieved by several ways, which may be classified as under :

Natural Control

In this method the control is going on independently and human being have no any relation with this. The different ways are as follows :

1. Climatological factors—temperature, rainfall, light, atmospheric pressure and air currents.
2. Physical factors—mountain, seas, rivers, lakes, desert and forest etc.
3. Natural enemies—parasites, predators including insects, birds, reptiles and mammals, cannibalism etc.

4. Natural diseases—may be caused by entomogenous fungi, bacteria, virus, protozoa and other micro-organisms.

Artificial or Applied Control

It is performed by the man and this may be of following type :

(A) Chemical control-Chemicals are used to check the population of insect.

   (a) Insecticides.

      (i) Stomach poison

      (ii) Contact poison

      (iii) Fumigants

   (b) Attractants;

   (c) Repellents and Antifeedants

   (d) Pheromones and Hormones

   (e) Chemosterilants

   (f) Auxiliary, synergistic or supplementary substances

(B) Mechanical and physical control

(C) Cultural control including use of resistant varieties

(D) Biological control

(E) Legal control

Natural control : All methods that destroy or check the infestation and which do not depend upon man for their success or continuance are collectively included in this method. The natural factors that check the insect population, their distribution and activities are as follows :

Climatic Factors

Climate is the most important natural force in keeping

insect population under control. It includes temperature, rainfall, sunshine, atmospheric pressure and air currents etc.

Temperature : It is the most important factor of the environment. Besides having a direct effect on the development and metabolism of the organisms which determines their abundance, it also affects them indirectly through food and other environmental factors such as moisture, movement of air and-atmospheric pressure etc. Temperature controls the development, fecundity, feeding, movement and disposal or distribution of insects. The insects are cold blooded animal, the body temperature of which depends more or less on that of the environment. A warm, moist environment is favourable for the development of most insects. Extremes of temperature are destructive toinsect life, the optimum temperature for most of the species range between 13 to 40° C and above this aestivation begins while below this hibernation takes place.

It is clear from this that they remain active only for a particular time and in most of the months they are forced to remain inactive by nature itself due to the temperature fluctuation. The honey-bees, however, live in an environment of their own creation and are able to 'air condition' their nests.

Rainfall and humidity : After temperature, the moisture is another important factor which affects the insects in several ways. The body composition of all insects, like other organisms, consists mostly water and they contain 80-90% water. Since water is an essential constituent of the cell protoplasm, therefore, all the processes inside the body require water. One of the fundamental necessities of all the insects is to maintain a proper balance between the moisture contents of their body and that of their environment. The activity of some insects is dependent on the advent of rainfall, *e.g.*, ants, termites, white grub, grasshoppers, hairy caterpillars etc. Rainfall influences the insects not only by the total amount of the area but also by its distribution in time. It affects the breeding, multiplication, and out break of pests, *e.g.*, the red pumpkin beetle does not lay eggs, until it comes across moist soil and general opposition takes place by this beetle only on days following irrigation or a heavy shower of rain. The adults of red hairy caterpillars *(Amsacta moori)* though fully formed inside the pupae do not emerge till there are one or two showers of rain in June or July.

Light : Sunlight is the greatest single source of energy for almost all biological systems and there are different rays *viz.* X-rays, gamma rays, ultraviolet rays and cosmic rays which directly affect the development, reproduction and growth of insects. The growth and moulting of insects are accelerated or retarded according to their requirements of light. The direct sunlight itself may kill the hidden stages of insects, therefore, they tried to hide themselves and become active in its absence. It also regulates some other activities of various insects. There is correlation between the activities of honey-bees and hours of sunshine. Most of the cock chaffer beetles and many moths pass the day in concealment. Dusk is the most usual time for the flight and copulation of moths; for the emergence of winged white ants etc. The number of hours of light in a day length (24 hrs.), termed photoperiodism, has considerable influence on the development of aphids. By reducing the photoperiod the production of sexual forms is induced. Thus each insect has its own light requirement and slight variation on this may affect their life processes.

Atmospheric pressure : The atmospheric pressure like other factors also affect the activities of insects. The phototropic insects are more active during periods of high barometric pressure and specially when the barometer is rising. The insects are much more affected in low pressure areas than the high atmospheric pressure areas. The activities of some insects are directly influenced by the pressure for example due to low atmospheric pressure the chances of rain increased resulting the emergence of ants with their eggs for their safety.

Air currents : The disposal of insects to a great extent depends upon the wind as they generally fly with the air current although some insects like mosquitoes habitually fly against it.

Many insect species are known to spread to new countries through air currents. Large number of aphids have been found after a strong wind and many of them are destroyed by falling in the seas and rivers etc. Thus air currents also play an important role in natural control of insects.

Physical Factors

Physical factors play important role in the distribution and abundance of insects. They include the mountains, seas, lakes and rivers, deserts, physical and chemical structure of soil and forests :

Mountain : Large mountains provide effective barriers to the insect spread as they are unable to cross them more over in high peaks, throughout the year ice is found due to which the insect can not crawl or cross them and ultimately their spread is checked.

Sea, lakes and rivers : Some species of insect do not possess the power of flight, hence they can not cross the seas, lakes and rivers and finally their spread is restricted.

Soil structure : The soil type and its fertility determine the plant and animal life which in turn tend to limit the number and type of species and their abundance. Similarly the physical and chemical composition of the soil may determine the presence or absence of soil inhabiting species, for example the white grub of groundnut prefer the sandy soil and it can not thrive in clay soil.

Deserts : The insect can not cross the large deserts and thus they act as barrier. Moreover, due to the absence of vegetation and different soil texture inhibit the spread of insects.

Natural Enemies

Every insect has its natural enemies in the nature and with the increase in the population of insects, the natural enemies begin to increase as they get larger food supply. The important natural enemies are insect parasites and predators, birds, mammals and reptiles.

Insect predators : They feed on the living insects and consume more than one individual during their life span. The important predatory insects are as follows :

Coccinellids : They prey on aphids and leaf hoppers.

Chrysopa sp. : Its larvae feed on aphids.

Mantis : Feed on different insects.

Syrphids : They also feed on many insects.

Parasitic insects : The parasitic insects pass their life on the other insect and kill them. The important parasite of some insects are given below :

Trichogramma minutum — On sugarcane borers.

Bracon gelechiae — On pyrilla

Apanteles flavipes — On lepidopterous larvae.

Other enemies : Besides, insect predators and parasites there are still other which prey on phytophagous insects, *e.g.*, frog, toad, reptiles, birds etc. Some mammals like rats, bears also prey on insects.

Cannibalism : The larvae of some insects feed on themselves and thus reducing the population, *e.g., Heliothis* larvae.

Natural diseases : Like other animals, the insects are also suffer from the diseases in great number and their population is checked is nature. The important micro-organisms which cause diseases are as follows:

(i) Protozoa

(ii) Bacteria

(iii) Fungi

(iv) Virus

(vi) Nematodes

The details of these micro-organism will be dealt in the chapter of biological control.

Artificial or applied control : Applied control includes those methods, under the control of man, which are necessary and used when harmful insects have not been held in check by natural agencies. The following factors are pre-requisites for the success of insect control :

(i) Exact taxonomy and morphology of the insect.

(ii) Life cycle, natural and seasonal history of the insect.

After having the understanding on above points we may know the damaging stages of insect, its habitat, nature and weak point in its life cycle, which will be helpful in devising the successful control strategy. Under applied control the following methods have been included :

(i) Mechanical and physical control

(ii) Cultural control

(iii) Legal control

(iv) Biological control

(v) Chemical control

Mechanical and Physical Control

The reduction or suppression of insect population by means of specially designed machines or devices or by manipulation of physical factors of the environment are included under this group of control.

Mechanical Methods

The mechanical control incorporates the following devices:

Hand collection of insects : Hand picking of large sized immature or mature stages of insects is the most ancient method and this may prove very effective in the collection of egg masses and larvae of such insects which lay eggs in clusters and feed gregariously, *e.g.*, sugarcane top borer, Pyrilla, hairy caterpillars, grubs of mustard saw fly and various stages of Epilachna beetle.

The regular hand picking of rolled leaves, harrowing the caterpillars of cotton leaf roller or the dropping withered top shoot buds and bolls of cotton attacked by spotted boll worm are some of the example.

Hand nets : Hand nets are usefully employed against the flying insects. They may be caught very easily during their migration when they are in large number, *e.g.*, grass hoppers, jassids, bugs, pyrilla and moths etc.

Mechanical barriers : Barriers are used for preventing movement of insect which migrate by moving from one field to another the following are the common practices which generally followed:

(a) *Tree banding :* Adhesive or sticky bands around the tree trunks are used against those insects which infest trees by climbing the trunks, e.g., Mango mealy bug.

(b) *Fencing :* It is possible to prevent invasion of a crop by migrating insects through the use of walls around the field so that may not go to another field.

(c) *Trenching :* The crawling herds of army worms may be stopped by constructing deep furrows, around fields against their marching ways. Trench barriers are also used to stop the locust hoppers.

(d) *Other barriers* : Collar of individual plant to protect from cutworms; paper or muslin bags over fruits to protect from fruit flies, fruit sucking moths and anar butterflies etc., complete wrapping of tree trunks with paper against the flat headed borers are other examples of the use of mechanical barriers.

Mechanical traps **:** Many injurious traps have been made to catch insects but a very large percentage of them fail to offer adequate crop protection.

(a) *Light traps* : Light traps are generally used against those insects which are phototrophic, *i.e.,* attracted towards light. For this purpose a strong light may be placed or hanged in the field above a container having kerosinised water. A large number of insects attracted to light will be killed in the kerosinised water. Now-a-days, specially designed light traps to collect different types of insects are available and may be procured according to type and nature of insect. Light traps are not used only to attract and destroy the insect but also as indicator of the activity of pests in the field so that timely control operations can be undertaken.

(b) *Adhesive traps* : Many adhesive traps are used to kill insects and for the purpose the pieces of bags or cloths are dipped in gum solution or in any other adhesive material having insecticide. These pieces are hanged on the trees in the vicinity of fields, the flying insects will come across and attached with the trap and finally die. Some times an attractant is also used in these adhesive traps to attract the insects.

(c) *Hopper dozers or other dozers* : Such type of traps are made for catching insects of a particular group and these may be handled by man, animal or by a tractor. Some commonly used hoppers are hopper catcher, aphidozers, fly traps, moth traps, maggot traps and electric traps etc.

Physical Methods

Physical method involves specially manipulation of change in temperature or humidity or employ radiant energy in some way to destroy pests :

Use of temperature : Artificial heating or cooling of stored products, the mills or factories where such products are processed, is a common method of insect control.

(a) *Low temperature :* Nearly all insects become inactive at 60-40°F. Few insects are not killed at these temperatures unless exposed to them for a considerable length of time. Generally insect undergo hibernation at 20 to 30°F. It is not certain that exposure to such temperature will kill the eggs of such species as the grain weevils. Practically no damage occur from insects at temperature below 40°F. Although low temperature is not as effective as high temperature in killing insects but storage of food products or clothing, the temperature at below or near freezing point will prevent all insects damage.

(b) *High temperature :* It has experimentally been shown that no insect can survive longer when exposed to temperature of 140-150°F. Most of the insects including those which attack stored grains can be killed within 3 hours exposure to temperature from 125-130°F. In many countries mills and large cultivators have equipped their buildings with enough heating pipes to enable them to raise the temperature 125-150°F for several hours and thus kill all insects in the buildings. In tropical countries like India where the sun is strongly radiant during the greater part of year, solar radiation can be most economically employed in the case of insect infested materials such as cotton seeds, grains, pulses, spices, nuts and woollen goods etc. Super heating with burning charcoal can also be practiced to raise temperature in godown to destroy most of the stored pests.

Humidity : Some crops can be protected from insect pests either by flooding or by irrigation, at proper time. The draining of marshy land and other standing water is the most effective method of destroying mosquitoes and house flies. Drying of grains (less than 10% moisture) before storage is a sound practice to prevent insect damage by stored gram pests.

Light : The use of radiant energy to control insects has been experimentally studied by a host of workers. The light has been utilized to attract many strongly phototrophic insects into traps from which they can not escape. In general insects appear to be strongly attracted at about 3650 A, hence ultraviolet lamps have been used in some countries to trap the flying moths. The phototrophic response declines sharply in the violet region at 4400 A, minor attraction may occur at 4920 A (blue to blue green) 5150 A (green) or 5550 A (yellow green) but the longer wave length of 5560 to 7000 A (yellow to red) are relatively unattractive.

Ionizing radiation or radio activity has also been used for insect control the radio activity consists of particles, B particles, neutronts, X-rays and r-rays. Of these r-rays are extensively used because at much lower dosages these rays are capable of sterilizing insects. This method of controlling insects (sterilization with r-rays) has been used by knipling Lindquest and Bushland for the control of screw-worm fly in U.S.A. in 1937. Cobalt-60 source of r-radiation was used by these workers to sterilize the pupae which developed into adults and mated normally. The males require comparatively lower dosage (2500 r-) than females (5000 *r-)* for sterilization. This technique was successfully applied in 1958 to an area of about 64000 sq. miles of Florida where after the release of nearly 3 billion sterile flies over a 17 months period, eradication of screw worm flies was achieved. This technique (sterile male technique), no doubt, is applicable for the control of other insects also but the following principles have been established for its successful augmentation :

(i) Mass rearing of the insect must be possible.

(ii) Adequate dispersal of released sterile males must take place.

(iii) Mating behaviour of males should not be impaired by sterilization, *i.e.*, mating should be normal.

(iv) Female should normally mate only ones or if more matings occur, the sperms of sterile males must compete with those of normal.

(v) Initial population of insect should be checked so that there is no competitions between sterile and non-sterile males.

(vi) The sterile males released in fields should not damage crops.

Sound : Little work has been done on the use of sound waves in insect control. Adams and co-workers (1953-54) have employed these waves produced by feeding grainery weevil larvae to find infestation Frings and co-workers (1948) have killed insects with very high intensity of sound. The male mosquitoes are attracted by special type of sound and made sterile by electronic flash gun in Canada.

Cultural Control

Cultural control may be defined as control of a pest by slight variation, introduction or suppression of farm practices which are normally adopted in the cultivation of a crop. These practices may lead to the control of insect pest either by directly affecting their growth and multiplication or by minimizing the chance of their attack on plant through careful manipulation or through increase' in the internal resistance of plant. They must usually be employed well in advance of time when damage by the pest starts. This is the cheapest and practical method of control as it does not employ an extra money to be spent.

In order to control insects by cultural practices, it is necessary that one should know the life history and habits of insects so that the weak point of their life cycle may be pin pointed. A cultural practice effective against one kind of insect species may not be useful against a closely related species because of the difference in their habits. The cultural operations should be followed at proper stage of development of an insect. The following are the important cultural practices employed for insect control :

Ploughing or cultivation of soil : Insects are largely affected by the texture of soil, its chemical composition, percentage of soil moisture, the temperature and other soil organisms. Therefore, the various methods of soil management have a definite effect upon many insects. Deep ploughing of fields is a good method of controlling many insects particularly those which pass some portion of their life cycle in the soil. This method is recommended for hairy caterpillars, cut worms, white grubs, sphinged caterpillars etc. Periodically, breaking up of field bunds is useful to destroy grasshopper's eggs. Similarly racking and hoeing of soil around cucurbit plants and fruit trees is effective for destroying under ground pupae of fruit flies and also the red pumpkin beetle affecting vines.

Sanitation or destruction of crop residues and weeds : Crop residues and other traces remaining in fields after harvest provide food or shelter to many insect pests. The destruction of all stables after harvest of crops like paddy, maize, jowar, sugarcane, brinjal, cotton, sweet potato, and cucurbits etc. would prevent the multiplication of borers and other pests which would otherwise go unnoticed and appear on crop in the following season. The weeds growing around the cultivated field also serve as the host of many insects like the spotted boll worm and cotton leaf roller develop on many weeds growing in and around orchards. Hence the removal and destruction of these weeds is the only known practical method to prevent multiplication of these pests. Many insects hide or hibernate under loose bark in deep or unhealthy branches of trees. Removal of such shelter by scraping, pruning or plastering would prevent the multiplication of several pests such as bark eating caterpillar and shoot-stem borer.

Variation in the time of planting and harvesting : The time of planting has a marked influence on the infestation of crops by some insects. By changing or carefully selecting the time when a crop is planted, we may avoid the egg laying period of a pest or get young plants well established before the infestation starts, *e.g.*, cucurbit vines sown early are better able to withstand the attack of red pumpkin beetle than those that are sown late and are still in the seedlings stage at the time of emergence of the beetle towards the end of February. Similarly, early sown cotton is found to be less liable to the attack of jassids than late sown cotton. This is also true with crops like mustard against aphid, linseed against gall midge and sesamum against jassids.

Crop rotation : Crop rotation is one of the most frequently recommended methods of insect control. If a similar crop is grown year after year on a large area the insects of that crop predominate as they are getting continuous food. But if the second crop is grown next year, the insect dependant on the first crop may not be able to live on the second. Therefore, the crop planted in the field which was not grown previous year, will at least temporarily escape damage from its specific pests. Among insects that injured cultivated crops the number of general feeders (polyphagous) is very small, those which feed on the plants of one family are numerous, and there are many that feed on only a few very closely related species. Therefore, much can be accomplished in preventing most insects from becoming serious pests in our fields, if a good rotation is practiced where a crop of one plant family follows to that of a different family. Crop rotations are generally more effective for insects those are restricted feeders, have limited powers of migration of sluggish habit and are slow breeder spending a relatively long time in the feeding stage.

The use of resistant varieties : Resistance of plant to insect attack may be defined as the relative amount of heritable qualities possessed by the plant which influence the ultimate degree of damage done by the insects. The intensity of resistant varies widely from crop to crop with associated insects. Some varieties or strains of cultivated plants are more or less resistant to some of the insects that attack them, hence the breeding of such resistant varieties of important crop plant has become a major weapon against insect attack. In last decade attempts have been made in abroad as well as in India to breed resistant varieties against variety of insect pests. But very little success has been achieved in this direction due to change in their feeding habit, existence of bio-types of particular insect and lack of complete resistance in the plant type. With the formation of various All India Coordinated Crop Improvement Projects suitable conditions have been created for under taking this type of work.

To develop resistant varieties studies were made on 100 insects and their host plants in U.S.A. and as a result of this only six varieties against hessian fly and European corn borer could be developed. During the last 30 years attempts were made in Punjab to evolve resistant variety of cotton against jassids but no success could be achieved. Even U4 variety of South-Africa which is resistant against the said pest failed to survive here. In our country concerted attempts are being made to develop the resistant varieties of sorghum, maize, paddy, cotton, lady's finger, cucurbits, mustard, sugarcane etc.

More than 10,000 varieties from the world germ plasm collection have been screened against sorghum shoot fly and 10,196 varieties were tested against stem borer out of which 26 lines were indentified showing relatively less damage by this insect. About 2000 varieties of cotton belonging to 4 groups namely, arboreum, herbaceum, hirsutum and barbadense have been screened against jassids, boll worms and stem weevil at many locations in India. Generally, these varieties introduced from USA are susceptible to jassid. However, some cambidian types like B. 1007 and Bandawar and crosses between these varieties are reported to be resistant to jassid. Out of the 364 lines of pea tested against pea-pod borer only 3 lines were noted promising. The classical examples of some resistant varieties against particular pest are given in the following table :

S.No.	*Crop*	*Varieties*	*Insect*
1.	Apple	Winter majetin	Wooly aphids
2.	Grape	American Varieties	Phylloxera
3.	Wheat	Pawnee, Ponca, Ottawa	Hessian-fly
4.	Jowar	Atlas, sargo	Chinch bug
5.	Alfalfa	Ladack	Pea-aphid
6.	Alfalfa	Cody Lohantan	Spotted alfalfa aphid
7.	Maize	Ks-2234	Grasshopper
8.	Maize	Ks-1859	Corn leaf aphid
9.	Barely	Dictoo	Green bug

Other cultural practices : The following practices are also used in this method of control to regulate the activities of some insect pests :

Mixed cropping : Due to sowing of mixed crop, the distance among the same kind of plants is increased, therefore, insects do not reach easily to them and thus resulting poor infestation. If jowar and bajra are sown mixed the attack of jowar stem borer is not so severe. Similarly, the sowing of wheat, gram, and mustard minimises the attack of termite, gram caterpillar and aphids on their respective hosts.

Some insect pests require two kinds of host to complete their seasonal cycle. This is particularly true with certain aphids in kumaun hills. Under such conditions the alternative host should be destroyed to check the multiplication of pest.

Only healthy and non-infested seeds or cuttings should be shown to check the carry over of the insects. This practice is particularly important in the case of sugarcane which may contain borer larvae and mealy bugs, cotton seeds which may have the pink boll worm larvae and cuttings of various garden plants which often contain mealy bugs and aphids etc.

Proper prunning and thinning, conservation and regulation of soil moisture, good preparation of seed bed and the judicious use of fertilizers offer possibilities of stimulating plant growth in such a way so as to grow profitable crop with minimum insect damage.

Removal of dry cane leaves to destroy the egg masses of pyrilla during winters and earthing of cane plants to prevent the attack of borer are some of the other useful cultural practices.

Sowing of trap crop is also another method to destroy and check the infestation of insect pests. This is particularly true with cotton when okra is grown to reduce the infestation of red cotton bug.

Legal Control

In the early days of agriculture development, the plants and their products were imported or exported with little or no thought to the insects or other pests that might be transported along with them. It is only in the middle of the 19th century when some countries thought about the legislation to restrict or prevent the spread of insects and other pests. The introduction of grape phylloxera from America to the grape-wine yards of France in 1860 caused such a serious destruction to the France vine that it was felt imperative to do something to prevent the unrestricted movement of infested grape vine to other part of the world. In 1881, the representatives of many European countries where grapes were extensively grown met and agreed to impose regulation restricting the movement of infested grape vines. Similarly the spread of sanjose scale in U.S.A. in later part of the 19th century laid to the passage of insect legislation in that country in 1905. In India, plant quarantine rules and regulations were made under the provision of the Destructive Pests, Plant Diseases and Noxious. Weeds Act No. 11 of 1914 in the guidance of Directorate of Plant Protection, Ministry of Food and Agriculture, Storage and Quarantine, Government of India. Before the enforcement of legal control following three important insects entered in the country :

***Icerya purchasi* :** It is a pest of citrus and came from Australia.

***Aspidiotus perniciosus* :** It attacks apple and pears and entered from China.

***Eriosoma lanigerum* :** It is also a pest of apple and pears and came from European countries.

The insect legislation may be broadly grouped into 5 classes:

(i) Legislation to prevent the introduction of new pests from foreign countries.

(ii) Legislation to prevent the spread of established pests within the country or within the state.

(iii) Legislation to enforce the application of control measures which have been found effective in preventing damage by established pests.

(iv) Legislation to prevent the adultration and misbranding of insecticides and to determine the impermissible residue tolerance limits in food stuff.

(v) Legislation to regulate the activities of pest control operations and the application of hazardous insecticides.

Quarantine laws : The quarantine laws are the regulatory measures dealing with the introduction and spread of injurious insects and other pests. In India this responsibility rests with the Central Government under the provision of Destructive Pest and Insect Act of 1914, with the following rules :

(i) Prohibiting or restricting the import of plants and plant materials, insects and fungi into India.

(ii) Prohibiting or restricting the movement of insects or diseases and their hosts from one state to another state in India.

Under these rules, plant and materials are to be imported through one of the following prescribed ports of country :

Sea ports : Mumbai, Kolkatta, Cochine, Dhanukushi, Kenkush and Bhava Nagar (only for American and West Indies cotton), Naga Patnam, Vishakhapatnam, Trivendram (only for unmanufactured tobacco), Chennai and Tuticorin. The ports of Chennai, Mumbai and Cochine have fumigating chamber's where fumigation is done.

Air ports : Amritsar (only for fruits and vegetables) Mumbai, Kolkatta, Chennai, Palam and Safdarjang Delhi and Trichurapalli.

Land frontiers : Khalra and Husainiwala for Pakistan and Guwahati for Burma and Bangla Desh.

All plants and plant materials are subjected to inspection and treatment, if necessary, at the port of entry. An official phytosanitary certificate in the International from the country of origin should accompany the consignment.

The plant quarantine department inspects the materials at three places :

(i) At the place from where material is being sent.

(ii) At the place where it reaches.

(iii) At the place where it is being kept.

The quarantine laws are so strictly followed that even beneficial insects may not be imported except at the following institutes :

(i) Indian Agriculture Research Institute, New Delhi.

(ii) Forest Research Institute, Dehradun.

(iii) Indian Veterinary Research Institute Mukteshwar.

(iv) Zoological Survey of India, Kolkata.

(v) Commonwealth Institute of Biological Control, Bangalore.

Complete restriction has been imposed on the import of following items as there is a chance to have such insects and diseases which are not found in India :

(i) Unginned Cotton, from Australia and Philippine, Newgni and Fijo islands.

(ii) Potato, onion, garlic and unmanufactured tobacco.

(iii) Unfumigated American Cotton.

(iv) Sugarcane, Seeds of Berseem and Sunflower etc.

Domestic quarantines exist against *lcerya purchasi,* bunchy top of banana, banana mosaic, sanjose scale and potato wart.

Insecticide laws : Before the enforcement of Insecticides Act, 1968, pesticides were included under the Indian Poison Act, 1919, and Drug Act, 1940. Thereafter the Food Adultration

Act, 1955 and Pesticide and Appliances Act, 1964 were enforced to regulate the standards of pesticides and appliances through Indian Standard Institute, New Delhi and Pest Control Committee. The standard materials are blended with ISI mark and these organisation also verify the information given on the labels pasted over the containers, equipments and appliances. In order to regulate the import, manufacture, sale, transportation, distribution and use of insecticides so as to prevent the risk to human beings or animals the following rules were framed in 1972 under Insecticide Act, 1968:

1. All the insecticides which are to be exported or manufactured must be registered at the central level. For this the registration form is to be submitted to the Insecticide Registration Committee at Directorate of Plant Protection and Quarantine, Faridabad. This application form along with the registration fee of Rs. 100/- through a chalan is submitted under the head 065. Other administration services, other receipt fees, fines etc.
2. Any State can have the licence for formulating or selling the insecticide only after registration.
3. The State Government appoint Licensing Officers who renew or issue the licence for the formulation and sale of the insecticides.
4. For the renew or grant of a new licence the following fee is deposited in the treasury through chalan under the head 105-Agriculture (Gha) other receipt (4) other items
   (a) For the formulation of the insecticide the minimum fees is rupees 50/- and maximum is Rs. 500/-.
   (b) For the sale, stock, demonstration and distribution of each insecticide the minimum and maximum fee are Rs. 20/- and 300/- respectively.

5. The Central or the State Government appoint insecticide inspectors in specified areas for the following duties :
   (a) To inspect the sale points of his area jurisdiction at least thrice in a year.
   (b) To make sure whether the provisions of this act or rules made thereunder are being complied or not.
   (c) To collect the samples of such insecticides which he feels substandard or do not meet out the requirments of the Act or rules made thereunder and are being sold in the market.
   (d) To verify the written complaints and to resolve them.
   (e) To exercise such other powers as may be necessary for carrying out the purpose of this Act or the rules made thereunder.
   (f) To maintain the records of the samples brought from the different sale points and to keep the inspection reports and to submit these reports to the officer concerned.
   (g) To inspect and enquire about the sale and use of insecticides which are against the rules of insecticide act.
6. For the inspection of insecticide formulations following specific duties have been assigned to the Inspectors :
   (a) To check the licences issued for insecticidal formulations in his area jurisdiction atleast thrice in a year and verify whether the terms and conditions mentioned in insecticide act are being followed or not.
   (b) To submit the report just after the inspection regarding the follow-up of rules and regulations concerning the insecticides and their formulations.

(c) To take the samples of any insecticide and send such samples to the Insecticide Analyst for test in the prescribed manner.

(d) To report the Government about the cases of insecticidal poisoning.

(e) Period of the validity of the licence :

  (i) The licence will be valid for two calendar years from the date of issue or its renewal,

  (ii) Application for the renewal of licence should be given before its expiry and the following amount will be charged after the date of expiry.

    (a) For the licence of insecticidal formulation Rs. 50/-.

    (b) For other licence Rs. 20/- only.

(f) *Appeal* : Any person aggrieved by a decision of licensing officer may appeal to such authority (Director Agriculture) in such manner or payment of Rs. 10/- by treasury chalan within a period of 30 days from the date on which the decision is communicated to him.

7. There is a provision for Insecticide Analyst and Laboratories at State and Central level under this Act.

8. Packing and labelling of insecticides as per rules has been prescribed in the Act.

9. Provision has also been made for precautionary facilities, *viz.*, clothes, gas masks, gloves etc. to the persons handling insecticides during transportation and storage.

Compulsory clean-up measures **:** All the state Governments have legislation for the control or eradication of insects, other pests, diseases and weeds. Provision under the act authorise the restriction or prohibition of insect, destruction of insects and other pests, diseases and weeds and the plants or plant materials which harbour them. Areas or crops can be declared under the provision of laws as "pest infested or disease infected". Rules are made operative from time to time to deal with serious situation of insect pests, diseases, and weeds. The operation of the 'State Insects Diseases and Noxious weeds Act is the responsibility of the government through Agriculture and Revenue departments. An Act namely "Cotton Pest Act" was enforced in 1919 in Madras under which it was made compulsory that all farmers must destroy the residues of old crop by the end of August I of each year. Similarly, with the enforcement of Baroda State Agriculture Pest and Disease Act, 1940, it was necessary to destroy the stubbles of cotton by the end of 25th May. The following were the rules made by different states in respect of compulsory clean up measures :

1. The coorg Agriculture Pest and Disease Act, 1973.
2. The Patiyala Destructive Insect and Pest Act, 1943.
3. The Bombay Agriculture Pest and Disease Act, 1947.
4. The Rewa State Agriculture Pest and Disease Act, 1944.
5. The East Punjab Agriculture Pest, Disease and Noxious Weed, Act, 1949.
6. The Assam Agriculture Pest and Disease Act, 1950.
7. The Uttar Pradesh Locust Destruction Act, 1951.
8. The Bihar Pyrilla Destruction Act, 1953.

The person living in these states are bound to follow the instructions mentioned in the Act and failing to do so they are punished as per rules mentioned therein.

Biological Control

The use of natural enemies for the control of harmful insects, other animals and plants is known as biological control. The term biological control was first used by Smith 1919 to signify the use of natural enemies. In other word collecting and rearing of natural enemies in great numbers in the laboratories and releasing them on harmful insects, other animals and plants is known as biological control. In general we can say that the study and utilization of predators, parasites and pathogens for the control of injurious insects is known as biological control. Following are the natural enemies which are used in this way :

1. Predatory and parasitic insects.
2. Predatory vertebrates birds. Macrobial control
   reptiles, mammals etc.
3. Parasitic nematodes
4. Protozoal diseases
5. Parasitic fungi Microbial control
6. Bacterial diseases
7. Viral diseases
8. Rickettsae

Following are the four principles employed in the biological control :

1. Collecting parasites, predators and pathogens from the places where they naturally assembled in great numbers and releasing them in places where they may do most good.
2. Collecting and handling the host insects in such a way so as to kill them but permit any parasites or predators among them to escape.
3. Rearing of predators and parasites in the laboratories in great numbers and releasing them in such places where the population of host is reached to its maximum.

4. Importation of natural species from other places (foreign countries) and use them against the pests established from the foreign country.

Basic Steps of Biological Control

Following are the steps which have to be followed for the successful biological control :

1. Exact determination of controlling insect species or race to be controlled.
2. Careful study of the geographical distribution of the pest species.
3. Knowledge about the biology of the host insects and their natural enemies, parasites attacking on different developmental stages of the host insect.
4. Thorough knowledge of the indigenous parasites and predators of the insect pests.

Careful study of the natural enemies that they are entomophagous and also ensure that they are predatory, parasitic, if not, they may not be used.

Attributes of Effective Natural Enemies

A good natural enemy may have the following qualities:

1. The most effective attribute of a natural enemy is its searching capacity of their host. When the host population is scarce it should be able to find the host.
2. It should be highly specific or monophagy but do not pass their life on other hosts.
3. High fecundity, short life cycle, so that they may be produced in large number in a short period. More generations of the enemies may be developed to the generation of the host.
4. Adoptability to new environmental conditions.
5. Resistant to pesticides.

6. Free from hyperparasites.
7. Power for dispersal at least as much as the host.

History of Biological Control

Since ancient times; Horticulturist of Asian countries is using predaceous ants for reducing the number of foliage feeding insects of horticulture crops. But insect parasite *Apanteles glomeratus* was observed attacking on cabbage butterfly *Pieris rapae* L. which was first described by Vallisniesi in the year 1661. After that little attention was paid in this direction. In 1800 Erashus Darvin suggested in his book Phytologia that the predators and parasites may be used for biological control. After which scientists started thinking in this direction and made successful experiments. The remarkable success in this direction was the control of cottony cushion scale *(Icerya purchasi)* by the import of a lady bird beetle *(Rodolia cardinalis)* from Australia in California (USA) in 1888. This lady bird beetle was used in 32 countries of the world where it established well and completely checked the population of cottony cushion scale *(Icerya purchasi Mash)*.

After that *Georgh Campier* 1890 made survey of different countries of the world and collected predators and parasites of fruit and scale insects and sent to America.

In 1906 a *Hawaiian Frediricik Mure* surveyed the natural enemies of sugarcane pests, and found Scolia *moniliae* was successfully controlling the *Anomala orientates*.

Agents of Biological Control

Vertebrate predators : In this category different types of animals found to catch and devour the insects, *viz.*, Fishes, frog, toad, tortoise, snakes, lizard, birds (Hudhud, Mama, Blue-Jay, Partridge, Quail, Peacock, Papiha, Owl, Bhujanga, Cuckoo, Crow, Sparrow and Kite) mammals (Bats, Rats, Mole and Mongoos). Among the predatory vertebrates birds are most important but they are less used to introduce from one country to another. Indian maina was introduced from India to Hawai island. It is clear from these examples that these natural enemies are doing as a autocidal control of insect in the nature but they are not being used as biological control agents.

Predatory and parasitic insects : They are also controlling the insects in the nature. By seeing their effectiveness in nature some natural enemies are reared in the laboratories and utilized for the biological control of insects. In America from 1883 to 1894, 403 natural enemies were introduced from other countries in which only 94 established in the country. In India also a large number of predators and parasites have been introduced from different parts of the world of the biological control of harmful insects and only 12 are successfully utilized.

Predators are those insects which catch and devour the harmful insects while parasite makes their homes on or inside the body of the host and pass their life cycle and finally destroy the host insect. Some of the predators and parasites are given below which have been successfully utilized in different parts of the world of the biological control of insects.

Predator/Parasite	***Host Insect***
Rodolia cardinalis (Predator)	Cottony cushion scale *(Icerya purchasi)*
Coccophagus gumeyi (Parasite)	Cirtophilus mealybug
Tetracnemus pretisus (Parasite)	*(Pseudococcus gahani)*
Anarhopus sydneyensis (Parasite)	Long-tailed mealybug
Tetracnemus perigrinus (Parasite)	*(Pseudococcon-adonidum)*
Pseudopycus utilis (Parasite) mealybug	*(Pseudococcus-nipae)*
Cyrtorhinus mundulus (Predator)	Sugarcane leaf hopper (*Perkinsiella* saccharicida)
Microceromasia sphcnophori (Parasite)	Sugarcane weevil *(Rhabdoenemis obscura)*

Following facts are observed from study and utilization of natural enemies in the biological control :

1. 50% insects controlled are belonging to the Homoptera (Coccids, mealybugs, aphids, jassids and white flies) and rest to the orders Coleoptera and lepidoptera.
2. Success of the biological control is depend on the dominant natural enemy.
3. Four times more insects are controlled by parasites in comparison to predators.
4. Biological control is most successful in temperate region in comparison to tropical and subtropical regions.

Natural Enemies Imported in India

The following 13 natural enemies are found successful which have been imported from different parts of the world in India to control the insects given against their name.

	Imported predator/Parasite	***Host insect/Host-animal***
1.	*Rodolia cardinalis (Predator)*	Cottony cushion scale *(Icerya purchasi)*
2.	*Aphelinus mali (Parasite)*	Wolly aphid *(Eriosoma-lanigerum)* potato tuber moth
3.	*Bracon gelechae (Parasite)*	Potato tuber moth *(Gnorimoschema opercuella)*
4.	*Prospaltella Perniciosi (Parasite)*	Sanjose scale
5.	*Aphytis mytilaspidis (Parasite)*	*Quadraspidiotus-Pemiciosus*
6.	*Spoggosia beniziana (Parasite)*	Coconut caterpillar *(Nephantis serinopa)*
7.	*Opius vandenboschi (Parasite)*	Fruitfly *(Dacus* spp.)
8.	*Apanteles flavipes (Parasite)*	Jowar stem borer *(Chilo partellus)*

Contd....

	Imported predator/Parasite	Host insect/Host-animal
9.	*Telonomus nawai (Parasite)*	Castor semilooper *(Achoea janata)*
10.	*Trichogramma minutum (Parasite)*	Sugarcane and cotton borers
11.	*Lixophaga diatraeae (Parasite)*	Sugarcane borers
12.	*Euglandina rosea (Predatory snail)*	Giant african snail *(Achetina fulica)*
13.	*Syngamia haemorrhoidalis*	Lantana weed

Qualities of Introduced Natural Enemies

1. Ability to control such pest for which it was introduced.
2. It will never be become phytophagous.
3. It will not destroy the indigenous primary parasite and do not become hyperparasite.

Use of micro organism in biological control : Very small and microscopic component used in biological control are put into micro organism. That's why it is called as microbial control.

Bacteria : de Herelle 1911-1916 gave the 1st practical biological control of insects on the use of bacterium *Coccobacillus acridiorum* against locusts. From that time bacteria are being utilized for the control of harmful insects. In the beginning of 1940 in the United States, it was observed that milky disease A and B in Japanese-beetle *(Popilliae japonica)* is caused by the bacteria *Bacillus popilliae and Bacillus lentimorbus*. These bacteria are successfully utilized for the control of Japanese beetle grubs. Infected larvae are crushed and mixed into the soil where they parasitise and kill the larvae. Now-a-days *Bacillus thuringiensis* is successfully used for the control of lepidopterous pests in different parts of the world. The various commercial formulations are available which are used to control the pests, *e.g.*, Dipel, thuricide, Biolip, Bio-tox etc.

Entomogenous bacteria may be classified as under :

- Spore forming
  - Obligate
    1. *Bacillus popillae*
    2. *Clostridium brevifaciens*
  - Facultative
    - Crystalliferous
      1. *Bacillus thuringiensis*
    - Non-crystalliferous
      - *Bacillus cereus*
- Non spore forming
  - Potential
    1. *Pseudomonas*
    2. *Aerobactor*
  - Facultative
    1. *Serratia marascens*

Fungi : In the nineteenth century it was thought that the fungi may be utilized for the control of insects. From that time the fungi are being used for the control of injurious insects. These fungi attack the host body externally or internally where the spores germinate and cover the whole body and kill the host. In America, white muscardine fungus *Beauveria bassiana* has been successfully used for the 1st time against chinch bug *(Blessus lucupterous). Spear and Colley* 1912 used the fungus *Entomophthora aulicae* for the successful control of brown tail moth. Other species of entomophthora are also used against other insects. *Empusa spp.* has also been used against houseflies and grasshoppers. All the spores of above fungi are mixed in water and sprayed on the crop. These fungal spores attach to body of host where they germinate and produce mycelium and hyphae and kill the host. High humidity is needed for the successful germination of the spores. Other fungi which are effective for the biological control of insects are *Isaria, Spororichum and Aspergillus* etc.

Virus : Although the important role of virus diseases as natural control factor has long been recognised. But, it is only in the last few years that the viruses have been used successfully in control programme. First successful establishment of an exotic virus disease was accomplished by Balch (1946), who introduced the polyhedrosis virus of the European spruce-saw fly *(Diprion heraynial)* into Newfound land. The first successful attempt to introduce a virus disease not present previously in the North American continent was made by Bird (1953) who applied a polyhedrosis virus of the European pine saw fly *(Neodiprion sertifer)* obtained from Sweden.

Now-a-days polyhedrosis virus *(Borrelina campeolis)* has been successfully utilized in California for the commercial control of alfalfa caterpillar by spraying of virus suspension containing SO lac polyhedra/c.c. @ 5 gallon/acre (Thompson and Steinhaus). Steinhaus also used the polyhedrosis virus successfully against *Helio this armigera and Heliothis virescens.*

Entomogenous viruses may be grouped into two categories inclusion viruses producing inclusion bodies in the host cell and non-inclusion viruses which do not do so. The inclusion viruses are further classified into polyhedroses which produce polyhederal bodies and granulosis viruses which produce granular bodies. Polyhedrosis could inhabit the nucleus and are called nuclear polyhedrosis viruses or the cytoplasm which are known as cyto plasmic polyhedrosis viruses. The inclusion viruses are more effective as insect control agents than the non-inclusion viruses.

Protozoa : There are so many parasitic protozoa as reported by *Tanada* 1959 "that the parasite protozoan's may be used successfully as a biological control agent because of their abundance" Due to various hinderances in their mass production not much work has been done on this aspect. It may be one of the causes that they produce cronic disease infection inspite of acute disease as a result of which the host insects may survive for a longer period.

In America *Perezia pyraustae* has been successfully utilized for the control of European corn borer *(Ostrinia nubilalis).*

Nematode : Some nematode species are also known which parasitize and destroy the insects., Glacer and Wilaox 1918 described in detail for using nematodes as a biological agents.

In 1959 Tanada also reported the presence of nematodes as international parasite on various beetles, grasshoppers, moths and other insects and recommended that they may be used successfully for the control of these insects. Now-a-days DD-136 *(Neoaplec-tana carpocapsae)* is used against the control of various insect pests. The nematode suspension is sprayed on the crops for the control of lepidopterous, Coleopterous and other insect species. In India it has been found successful against a number of pests. The other nematode species are Neotylenchus, Rhabditis, Mermis, Hercamermis Protrellus, Pseudonymus Sternienonema etc.

Problems of Biological Control

1. Biological control agents have narrow host range and control only a particular pest.
2. The effect is very slow and required a long time for the control.
3. Highly trained personals are required because living agents are used.
4. More member of biological control agents are required to use so that much money and time required.
5. Some times introduced agents are also parasitized by parasites that's why no success achieved.
6. Some agents may have special environmental conditions for its development and multiplications, therefore, require waiting period.
7. Due to lack of suitable environmental conditions the introduced parasites or predators do not adopt the prevailing conditions. More time is required to study the adoptability of that agent.

Scope of Biological Control

1. Using agents are no toxic and do not leave any harmful residues.
2. Effect of biological control is permanent type. Once introduced may work constantly in the nature for several years.
3. They do not harm to beneficial organism as like chemical control.
4. Some agents can easily be mass produced economically and may be utilized.
5. It is free from resistant problem as like insecticidal resistance.
6. Some agents may be used in combination with insecticides. If insect save from the attack of insecticide may be controlled by the natural enemies.

Chemical Control

The chemical which either kill the insects or repel them are used in chemical control. Insecticides are included under this group of control and are those substances which are used for protecting plants from harmful insects. The chemicals that kill the pests (any living organism which is harmful) are known as pesticides which may be classified as under.

S. No.	*Pesticides*	*Against which used*
1.	Insecticides	Insects
2.	Nematicides	Nematodes
3.	Acaricides	Mites
4.	Rodenticides	Rodents
5.	Molluscicides	Snails and slugs
6.	Fungicides	Fungal diseases
7.	Weedicides	Weeds

Insecticides

The detailed classification of insecticides has been shown in the chart. On the basis of entry in the insect body they may be grouped as follows :

1. Stomach poison
2. Contact poison
3. Fumigants

Similarly on the basis of their origin they may be classified as :

1. Inorganic compounds
2. Organic compounds of plant origin
3. Synthetic organic compounds

Inorganic Insecticides

These compounds were mainly used before the introduction of DDT and other synthetic organic compounds. Their use now is greatly restricted, and that too only for the control of some specific pests. The following compounds have been included in this category :

1. Arsenic compounds
2. Fluorine compounds
3. Sulphur and its compounds
4. Phosphorus compounds
5. Antimani compounds
6. Boron compounds
7. Mercury compounds
8. Thallium compounds

Arsenic compounds : The various compounds of the element arsenic are the most important and widely used stomach poisons for insects. Some of these arsenicals meet nearly all the requirements for an ideal stomach poisons, except that they leave residue on the treated plants which may be dangerous to man and animals. The commonly used arsenicals are as follows :.

Lead Arsenate

It was first used as insecticide in U.S.A. in 1892 for controlling the gypsy moth. It is generally used in two forms:

1. Acid orthoarsenate
2. Basic orthoarsenate

The commercial lead arsenate is a mixture of these two in which acid form predominates. Lead arsenate is relatively weak poison as compared to other arsenicals. It is unsoluble in water, therefore, possesses high degree safety to foliage, one of the advantages in using lead arsenate is that it only affects insects feeding on treated material, while other useful insects like parasites and predators remain unaffected. Lead arsenate may be applied either as dust or sprary, but more generally used as a spray. As a dust, it is generally diluted with from 2 to 20 parts of some inert carriers such as talc, hydrated lime or gypsum. As a spray 1.5 to 2 kg of lead arsenate is dissolved in 500 lit of water which will be sufficient for about a hectare. However, different concentrations have been recommended to various pests, *e.g.*, 0.25% spray for the control of lemon butterfly, epilachna beetle and red pumpkin beetle, 0.6% spray for the control of castor semilooper. An equal quantity of hydrated lime is added to the insecticide to prevent phytotoxicity to tender foliage.

Calcium Arsenate

It was used as insecticide as early as in 1907 but the compound contains so much water soluble arsenic that it burnt the foliage badly. Later, it was used in 1920. The commercial calcium arsenate are a mixture of tri-ca-arsenate, acid ca-arsenate with a considerable excess of lime and calcium carbonate. The safe calcium arsenate have basic calcium arsenate as a principal constituent which hydrolyse much more slowly. Calcium arsenate may be applied in the same way as lead arsenate but it is mainly as a dust and rarely as spray or in bait material because of relatively poor adhesive properties. It is cheaper than lead arsenate but more toxic than the former to the insects. The old calcium arsenate is more toxic to plants also and only resistant plants such as cotton, potato and apple withstand it. The major demerits of its are as follows:

1. It is not more stable.
2. The danger of burning foliage.
3. The toxicity to higher animals.
4. Its incompatibility with nicotine sulphate, soap, sulphur, pyrethrum and chlorinated hydro-carbon.

Paris Green

This was the first important stomach poison and was developed in U.S.A. in about 1865 to control the colorado-potato beetle. It is a acito arsenite of copper and is green coarse powder on account of its high arsenic content it is a quickacting insect poison but due to its great solubility in water, it may cause a severe scorching of the foliage. This insecticide is no longer used in India due to its high toxicity. It has now only a limited use as a mosquito larvaecide and for the control of snails.

Sodium Arsenate

This is very highly soluble form of arsenical, therefore, is mainly used in poison baits. It is extremely toxic to plants, therefore, extensively used as a weed killer. In the preparation of baits for the control of locust and grasshoppers, it is used with wheat bran in 2% ratio.

Basic Copper Arsenate

This compound has been developed comparatively recently as a safe and stable arsenical for foliage application. It had advantage of being quite insoluble and resistant to hydrolysis by water and is only slightly affected by exposure to carbon-dioxide. It is about as toxic as lead arsenate to a variety of insects.

Fluorine compounds : These are derivatives of hydro-fluoric acid (sodium fluorite), Fluo-silicic acid (sodium fluo silicate, barium fluosilicate) and fluo-aminic acid (sodium fluo aluminate). The following are the main characteristics of these compounds :

1. They kill the insect more quickly than the arsenicals.
2. They are cheaper.
3. Less toxic to mammals.
4. Comparatively safer to plants.
5. They act both as stomach as well as contact poison.
6. Some cases they have repellent action also.

Sodium Fluosilicate (Na Si F)

It is a white granular, odourless powder containing 60.6% fluorine. This compound is also soluble in water, therefore, should not be sprayed on plants. It is widely used in poison baits to control grasshoppers, locust cutworms and crickets. Since better and more effective insecticides are available hence this insecticide is not much used now.

Barium Fluosilicate (Ba Si F)

Barium fluosilicate contains 40.8% fluorine and is least soluble of the inorganic fluorine, therefore, can be used in the form of spray. It is used at the rate of 1.5 to 3 kg barium fluosilicate in 500 litres of water per acre to control the flea beetles and blister beetles. Dust formulation may be prepared by mixing 1 part of it to 3 parts of inert dust viz. chalk, talc, clay or tobacco dust. It is reported to be corrosive to parts of spray machinery. It is incompatible with nicotine, calcium arsenate, lime sulphur, bordeaux mixture and alkaline soaps.

Cryolite or Sodium Fluo Aluminate (Na 3 Al F 6)

It is found in a natural form as a mineral in green land. In the mineral form, it is crystalline and contains 90 to 98% sodium fluoraluminate. The synthetic cryolite is lighter and fluffer than natural cryolite and contains 86% Na 3 Al F6. It is less soluble in water but soluble in dilute acid and alkalies. It is used as a spray at the rate of 2.5 kg in 459 lit of water per hectare. For dust preparation it is diluted with inert diluents such as clays and talc in the ratio of 1 : 2 and used many chewing insects.

Sulphur and its compounds : Sulphur is used against insects, mites and fungal diseases. Sulphur acts on fungi both by contact and by producing gases which have fungicidal action. It has little residual action, but due to following reason it-is not used frequently.

1. Due to its odour.
2. Corrosive action on metals.
3. Bad effect on the colour of cloths.
4. Germination power of seeds is reduced.

It is used in the form of dust, wettable powder and spray as lime sulphur.

Lime Sulphur

Lime Sulphur is prepared as follows :

1. Sulphur	2 kg
2. Lime	10 kg
3. Water	50 lit

The lime is dissolved in big container having 1/3 of water (*i.e.,* about 16-17 lit). The sulphur is mixed gradually. This mixture is boiled for about an hour and remaining water is added. When the colour becomes brown it is allowed to cool. The mixture should not be heated more otherwise it will be highly phyto-toxic. 40 lit of this mixture may be mixed with 1000 lit of water to spray the solution in a hectare.

Lime sulphur may also be used in the form of the dust which may be prepared by mixing 1 kg sulphur in 5 kg hydrated lime this is used @ 25-30 kg/hect.

Phosphorus compounds **:** Among the phosphorus compounds zinc phosphide is commonly used against grasshoppers and cricket. For the control of rats it is used in the bait.

Antimony compounds (Tartar amatic) **:** It contains antimonile acetate and used as bait poison in combination with sugar as attractant. Tartar amatic is also used to prepare poison baits against fruit flies and thrips.

Tartar acid	5 kg
Mollasses	25 kg
Water	1200 lit

The above quantity is sufficient for a hectare.

Mercuric compounds (Mercuric chloride) **:** It is so called as mercurous chloride or corrosive supplement or calomel. This compound is not in much use at present but was previously used against root affecting insect such as larvae of flea beetle, cabbage maggots and onion maggots etc.

Thallium compounds (Thallium sulphate) **:** It is used with some sweet as bait carrier to form a valuable poison against ants.

Boron compounds **:** Amongst the boron compounds, Borex or sodium tetra borate is used to kill the maggots of house fly and mosquitoes in ponds and pits of farm yard manure.

Organic Insecticides of Plant Origin : The insecticides of plant origin extracted from seed, flower, leaves, stem and roots are termed as botanical insecticides. Unlike synthetic organic insecticides they are safer to use, but since they are expensive and lack residual toxicity, their use has been limited in India.

Nicotine : Nicotine occurs in atleast 15 species of *Nicotiana* but the commercially important species are *Nicotiana tobacum* and *N. rustica*. Nicotine commonly occur in *N. tobacum* plant to the extent of 2.5% in the leaves and 5 to 14% in *N. rustica*.

Tobacco was one of the first material to used as an insecticide as early as 1763. The use of finely powder tobacco mix with water was recommended to destroy aphids without injury to the foliage. In 1828, an alkaloid namely nicotine was recognised which was synthesised by Picted and Retschy in 1904. The alkaloid nicotine when freshly distilled is a colourless nearly odourless liquid boiling point 247°C, specific gravity 1.009 at 20°C and vapour pressure 0.0425 mm Hg. at 25°C. It can be purified by vacuum distillation. When exposed to air, its darkness changes from red to brown to black becomes more viscous and develops specific odour. It is soluble in alcohol, ether or petroleum ether and is miscible with water due to of its basic nature. Nicotine usually forms salts with acids and dibasic salt with many metals and acids. The nicotine sulphate $(C_{10}H_{10}N_2)_2 . H_2SO_4$ is widely used as an insecticide as it is less toxic to warm blooded animals than nicotine and is less volatile, therefore, more stable. When alkaline water or soap solution is added to nicotine sulphate, the free nicotine is liberated and shows higher insecticidal activity than the sulphate.

The nicotine enters in the body through the spiracles and acts on the ganglia of the nervous system. The less amount of nicotine produces excitement on the insect while more amount produces paralysis due to which death takes place. It can be used by various ways as follows :

(i) *As a dust* : The concentrated nicotine extract is mixed with talc, gypsum and sulphur to prepare the dust. The type of carrier has an important effect upon the dust.

(ii) *As a liquid extract* : A decoction of tobacco leaves with soap is prepared by boiling 1 kg of leaves, stalks etc. of tobacco in 10 lit of water. One kg of soft soap is added after the water has boiled. This is continuously stirred and after a thick dark coloured liquid is obtained, strained. This stock solution should be diluted 10 to 15 times with water before spraying.

A standardized 40% nicotine sulphate is also available to use as spray which remains as toxic residue for a short time after application. The nicotine sprays are diluted so as to contain 0.05 to 0.06% nicotine. It has been successfully used in the control of sucking insects, *viz.*, aphids, thrips, psyllids, leaf miners and jassids.

(iii) *As fumigants* : It is also used as fumigant either by vapourising the concentrated extract with heat or by burning papers or powders soaked in the extract.

Nicotine is compatible with many of the common insecticides and fungicides, including sulphur, lime sulphur and copper preparations.

Pyrethrum : The use of Pyrethrum as an insecticide originated in trans Caucasus region of Asia about the year 1800. For many years its nature was kept secret by these countries and the product was sold as flea and louce powder. Before IInd world war it was largely prepared in Japan but at present Kenya is leading in its production.

Pyrethrum occurs only the plants belonging to the genus Chrysanthemum = (Pyrethrum) family composite. The two species namely C. cinerariaefolium and C. coccineum possess a sufficient amount of toxic content. The former species is, however, of commercial importance. The active ingredients are :

1. Pyrethrin I, ester of pyrethrolone with chrysanthemum monocaroxylic acid.
2. Cinerin I, ester of cinerolone with chrysanthemum monocarboxylic acid.
3. Pyrethrin II, ester of pyrethrolone with chrysanthemum dicarboxylic acid monomethyl ester.
4. Cinerin II, ester of cinerolone with chrysanthemum dicarboxylic acid monomethyl ester.

The pyrethrin content of Indian flowers is 0.5% to 1.41% and inspite of 79 tonnes of flower production in Kashmir valley, 200-300 tons of flowers from Kenya which have 1.43-1. 83% pyrethrins are imported every year to meet indigenous requirement of pyrethrins. The cinerin is more stable than the pyrethrins and pyrethrin I and cinerin I seem to be more toxic than pyrethrin II and cinerin II, respectively.

Pyrethrum concentrates are prepared by ground flowers with solvents such as petroleum-ether, glacial acitic acid, methyl-alcohol or acetone. The pyrethrins are highly unstable in the presence of light, moisture and air. Whole flowers decompose more slowly than ground flowers or dusts. Pyrethrum preparations are rapid-acting contact insecticides, known for a quick knock-down effect on the insects. There is very limited residual effect and to enhance its toxicity and residual action, certain chemicals such as piperonyl butoxide are mixed to it. Pyrethrins are decomposed in alkaline solutions, therefore, they are incompatible with the lime sulphur, lime, calcium arsenate and paris green etc. Following preparations are generally available for use :

(i) *Dust* : It is prepared by mixing pyrethrin extracts with appropriate clay or talc carriers and formulated as 0.25 to 0.5% dust which may be used @ 25-30 kg/hect.

(ii) *Emulsifiable concentrate* : Emulsifiable concentrate containing 3 to 20% pyrethrins is formulated which may be used as spray by diluting in water as 0.002 to 0.004%.

(iii) *Aerosole bomb* : For use in aerosole bomb a 20% pyrethrin concentrate is used in such a way to give final concentration of 0.04 to 0.25% pyrethrins. It is used to kill the flying insects in the air.

Rotenone : Plants containing rotenoids were used as fish poison by the natives in tropical and some semitropical countries. But it was only in 1902, when rotenone was extracted from the roots of *Derris elliptica* and *D. malaccensis* in America. Rotenone has been extracted from the 68 species of plants including 21 *species* of *Teprosia*, 12 of *Derris*, 12 of *Lenchocarpus*, 10 of *Millattia* and several of *Mundulea* but two species of *Derris* are the economic one for commercial production. The rotenone content in the roots may vary from 4 to 11% depending on the variety. It is effective as a contact poison like pyrethrum, and can, therefore, be used against aphids, thrips, flea beetles and caterpillars etc., on vegetable crops. When exposed to light and air, it decomposes changing from colourless to yellow to deep red and resulting in non-insecticidal product. Rotenone is readily oxidized in the presence of alkali, hence incompatible with alkaline dusts and other alkaline wetting and spreading agents. Pure crystalline rotenone is prepared by extracting powdered rotenone containing roots with a solvent such as eather or carbon tetra chloride. It is harmless to human beings and animals but is extremely poisonous to fishes. It is soluble in ethyl alcohol, benzene, acetone, xylene, carbon tetra chloride and insoluble in water. It is formulated as below :

(i) *Dust* : Dust formulations are usually prepared by mixing dried ground roots with 3-5 parts of suitable carriers such as talc, clay, gypsum and sulphur etc. A dust containing 0.5% to 1.0% rotenone is generally used to control the insect pests.

(ii) *Wettable powder* : Wettable or dispersible powder may be made from finally ground derris roots by mixing with natural soap 5% wettable powder is generally available, 2 kg of which is mixed in 500 lit of water to control the insects.

(iii) *Extracts of derris* : These are available as dried reagents containing 25 to 30% rotenone and the rest amount of refined petroleum oils. Since rotenone has limited solubility at petroleum oils, natural solvents are generally employed to increase the solubility. The spray is used at a concentration of 0.001 to 0.002%.

Ryania

The root and stem of a South America plant *Ryania speciosa* have been found to have 0.16 to 0.2% insecticidal components, the most important of which is the alkaloid ryanodine. It is effective both as stomach and contact poison and soluble in water and most of the common organic solvents except benzene and petroleum oils. The ryanodine is more stable to the action of air and light than pyrethrum or rotenone and has considerable residual action which appears to be less toxic to mammals than rotenone. It is used as 20 to 40% dust and 0.3 to 0.6% spray to control a variety of pest. In foreign countries, Ryania is being used against borers of corn and sugarcane with benefit, but in India since it is imported and twice as costly as pyrethrum, hence its use is not so popular. Ld 50 = 750 mg/kg.

Hellebore

It is extracted from the roots of *Veratrum album* and *V. viride* which are incorrectly called as white and green hellebore, respectively.

V. *album* is native of Europe and Northern Asia while V. *viride* is found in U.S.A. The hellebore acts as stomach poison and specially used as larvaecide for house flies in manure. It is not in much use because of high cost, lack of standardization of toxic content and because only fresh ground material is effective.

Sabadilla

It is extracted from the seed of sabadilla plant *(Schoenocaulon officinale)* of South and Central America. The people of these countries have been using this as louse powder since long, but it was not until 1938 that a thorough investigation of their insecticidal properties were begin. It is present in the seeds upto 2-4% from which about 13% cevadine, ($C_{32}H_{49}O_9N$) 10% veratradine ($C_{36}H_{51}O_{11}N$) and lesser amount of civadilline, sobitine and sabidine have been isolated. Among these cevadine and veratradine are more insecticidally active and act as both contact, and stomach poisons. It is sensitive to light, therefore, should be kept in closed container in dark places.

Sabadilla is generally used as 10-20% dust or spray available in the form of 50% wettable powder. Its rapid deterioration upon exposure to light makes its use safer specially on crop at the time of harvest.

Quassia

It is found in the bark and woods of two trees namely *Quassia amara* and *Picrasma excelsa* from which Quassia and Picrasmin alkaloids, having same imperical formula ($C_{22}H_{30}O_6$) were identified. Decoction of Quassia in water with soft soaps have been found effective for the control of aphids and saw flies.

Scabrin

It is extracted from the roots of two weeds namely, *Heliopsis scabra* and H. *longipis* It is just like pyrethrum in nature and mostly used to kill the house fly.

Neem Seeds

The seeds of neem *(Azadirachta indica)* possess following four ingredients which have insecticidal properties :

(i) Azadirachtin

(ii) Nimbidine

(iii) Meliantriol

(iv) Nimbine

Suspension of powdered seeds at 0.1% concentration act as repellent for desert locust but the powdered material in the ratio of 1: 100 serves as protectant of grains against storage pest.

Anonaine

Powdered dried seeds, leaves, bark and roots of custard apple *(Anona reticulata, A. sguamesa)* have been found to exhibit insecticidal properties. The alkaloid has been identified as anonaine which acts both as stomach and contact poisons against insect pest.

Acorin

The roots of the plant sweet flag *(Acorus calamus)* produces acorin glucocide which has insecticidal properties. The insecticidal property of acorin has not been exploited fully but still the farmers of Orissa mix the powdered root with paddy for protection against storage pest.

Karanjin

It is furo-coumarin compound isolated from the acetone soluble fricton of the seed of Karanj *(Pongamia glabra)* which has manifested insecticidal property against various stored grain pests namely *(Tribolium castaneum, Sitophilus oryzae, Callosobruchus chinensis* etc.) Recently Karanjin has also exhibit, juveno mimmelic property against Indian Flesh fly.

Ricinin

The leaves of *Ricinus communis* contain the alkaloid ricinin which has been found toxic to the larvae of coddling moth.

Oils and Emulsions

The term oil is generic expression for a great variety of chemical substances which are composed of carbon, hydrogen, At ordinary temperature, they are greatly fluid insoluble in water, readily soluble in ether and chloroform etc. They are readily inflammable, slightly lighter than water and have peculiar property of penetrating into the pore of dry substances. There are mainly three classes of oils :

Fixed oils : These are the animals fat and vegetable oils such as fish oil, linseed, soyabean, castor and coconut oil. They are essentially glycocides which saponify or form soap with alkaline bases. Fish oil and soyabean have been mostly used for making insecticidal soap. The soap may be prepared with the following :

Fish or Mahua oil	2 kg
Rosin	3 kg
Castic soda	650 g
Water	3.5 lit

1. kg soap after dissolving in 25 litres of water may be sprayed on a tree to control the nymphs of *Drosicha mangiferae*. For the control of sugarcane white fly 5 kg of soap should be mixed in 1000 litres of water and sprayed in a hectare.

Aromatic oils : These differ from the fixed oils as being not greasy or viscous and not being capable of saponification. They derived from the plants and have a very pungent odour, e.g., citronella, camphene, menthal, piperment, euginol. Euginol is used to attract the fruit flies where as citronella to repel the mosquitoes.

Petroleum oils : These are derived from sedimentary rocks and are complex solution of hundreds of hydrocarbons from which are derived natural gas, kerosine, gaseline, lubricating oil, asphatar and other mixture. As per origin crude petroleum oils are usually classified as paraphenic base, nephthenic or asphalitic base and mixed base oils. Actually all crude petroleum oils are mixture of all three types of components. Petroleum oils have been proved good contact poisons but are dangerous to green foliage; therefore, rarely used on plants. Light oils such as kerosine, dutrex or aromax are often used as solvents or diluents of insecticides. Petrol is popular for the control of stem boress of mango, apple and orange. Since the petroleum oils are harmful to plants, therefore, their emulsion is used. In the form of emulsion, oils are commonly used in combination with soap which acts as emulsifier. Diesel oil emulsion is one of the commonest emulsion used in India against san jose scale, which may be prepared as under :

Diesel oil	10 kg
Soft soap	4 kg
Water	28 litres

Diesel is generally added to the boiling soap water solution. The emulation thus formed is diluted with 30 parts of water before being applied to the trees.

Synthetic Organic Insecticides

Perhaps the 1st use of synthetic chemical was in 1892 when the potassium salt of 4, 6-dinitro-ortho-cresol was marketed in Germany as insecticide. However, it was the development of DDT during the IInd world war which considerably proved that a synthetic compound would proved superior to a variety of inorganic and natural products for many insecticidal uses.

Since then a large number of chemicals have been produced as effective insecticides. Synthetic organic insecticides may be grouped as follows :

1. Dinitro-phenols
2. Organic thiocynates
3. Chlorinated hydrocarbons
4. Organic phosphates
5. Carbonates
6. Synthetic pyrethroides

Dinitro-phenols : The earlier use of the dinitro phenols as insecticide was in 1892 and they are the derivatives of 4, 6-dinitro 2-alkylphenols and their salts or esters. These are generally used against scale insects, aphids, ticks and mites and include following insecticides :

DNOC : It is yellow crystalline solid (mp. 85°C) soluble in water to 0.014% at 15°C and forms salt with organic and inorganic bases. 4, 6-dinitro-ortho-cresol has the following structural formula :

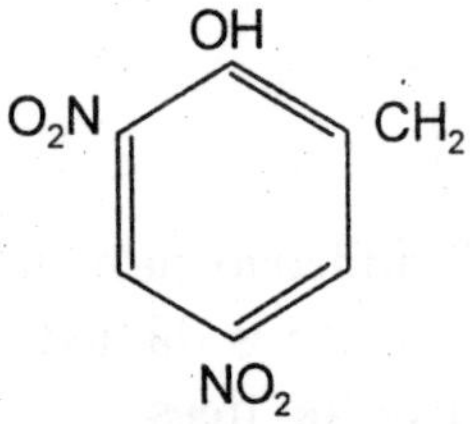

The compound readily forms salt with Na, K, NH, C and Ba which are water soluble. DNOC is a powerful insecticide, ovicide and acaricide but its pytotoxicity has restricted its use to the control of grass-hoppers, locusts and over-wintering insects in the form of baits. In India 0.1% of its Na salt has proved effective in the control of sanjose scale in Himachal Pradesh. It is also used as weedicide. The acute oral LD 50 for rat is 40 to 65 mg/kg.

DNOCHP ($C_{18}H_{24}N_2O_6$) : It is name given to 4, 6-dinitro-O-cyclohexylphenol. DNOCHP is yellowish white odourless crystalline solid, mp 106°C and only slightly soluble in water and solvents such as acetone, benzene, CCl_4, ethyl-acetate, ethylenedichloride, xylene, etc. Its structural formula is :

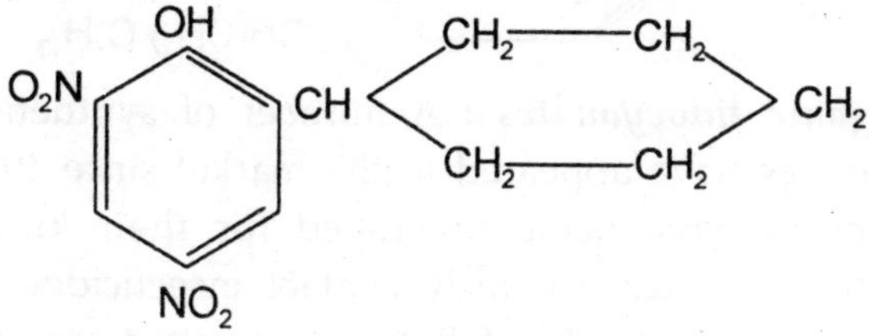

It is formulated as 20% wettable powder or 1.5% dust and mostly used for the control of red spider mites on fruit trees. It should be used with care upon all the plant foliage because of the injury which may result from over dose or unusual weather conditions of high temperature and humidity.

Both the above chemicals inhibit oxidative phosphorylation in insects and mammals and cause increased oxygen consumption which is responsible for toxic action of the compounds.

DNOSBP : It is 4-6dintro-O-sec-butylphenol and is crystalline orange solid, MP 42°C. A 36% solution of its salt in water is marketed and used as an effective dormant insecticide for orchard application. Its structural formula is :

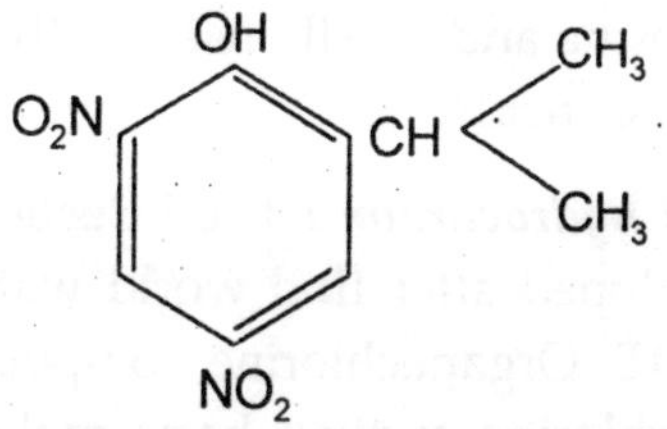

DINOCAP : It is 2-(l-methyl-n-hepty)-4, 6-dinitrophenyl crotonate or 4, 6-dinitro-2-caprylphenyl crotonate. DINOCAP is commonly known as karathane or crotothane and used as an acaricide and fungicide available as 25% wettable powder. It has a low toxicity to mammals. Its oral acute LD 50 for rat is 980 to 1190 mg/kg. The structural formula is given below:

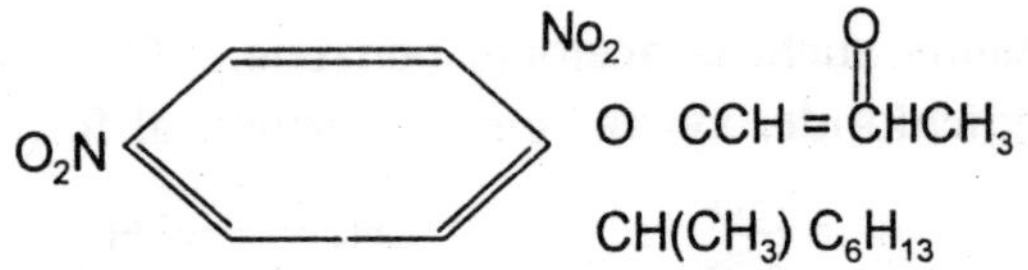

Organic thiocyanates : A number of synthetic organic thiocyanates have appeared in the market since 1932. These insecticides have been developed for their knock-down properties and are primarily contact insecticides with low mammalian toxicity. The following two products are mainly used under this group.

Lethanes : Louryl thiocyanate (C_{12} H_{25} SCN) is the main constituent of these thiocyanates. There are various trade names under which the lethanes have been formulated, *viz.*, lethane-60, lethane 384, lethane A-70, lethane B-72 and loro. It is generally used for the control of aphids, aleyrodids, thrips, spider and mites etc.

Thanites : The thanites are produced by treating the camphene from palm oil or tarpint oil with chloroacetic acid and than with an alkayl thiocyanate. The technical product containing 80% isobornyl thiocyanoacetate and the remainder closely related bornyl and fenchyl thiocyanoacetates. It is marketed as a technical product as well as 20% solution in deodorised kerosine and mostly used as fly, cattle sprays and as household insecticide.

Chlorinated hydrocarbon : The insecticides of this group have been developed after IInd world war and appeared in public with DDT. Organochlorine compounds have carbon, hydrogen and chlorine as their basic molecular constituents and some compounds may have oxygen and sulphur also. They may be grouped as under :

(i) Diphenyl methane structure — DDT and its analogues

(ii) Monocyclic structure — BHC

(iii) Bicyclic structure — Toxaphene, Strobane

(*iv*) Polycyclic structure — Aldrin, Chordane, Heptachlor

(*v*) Heterocyclic structure — Endrin, Dieldrin, Thiodan, Telodrin

Diphenyl Methane Structure

These compounds have few benzene rings attached to a single carbon atom and include DDT and its analogues.

DDT

It is dichlorodiphenyl trichloroethane (C_{14} H_9 Cl_5) and was first synthesized by German chemist Zeidler in 1874, but its insecticidal properties were not discovered until the work of Paul Muller in 1939. It is produced by reacting chloral hydrate with monochlorobenzene in the presence of concentrated sulphuric acid. Technical DDT is a white to cream coloured amorphous powder and consist of a mixture of atleast 14 chemical compounds. These included, 65-80% of active pp-DDT, 15-21% of the nearly inactive op'- DDT, up to 4% of DDD or TDE and other related materials. The principal ingredient of technical DDT, pp'-DDT, is 1.11-trichloro-2, 2 di-(4-chlorophenyl) ethane or I, I, I trichloro 2, 2-bis (p-chlorophenyl) ethane. The structural formula is as follows :

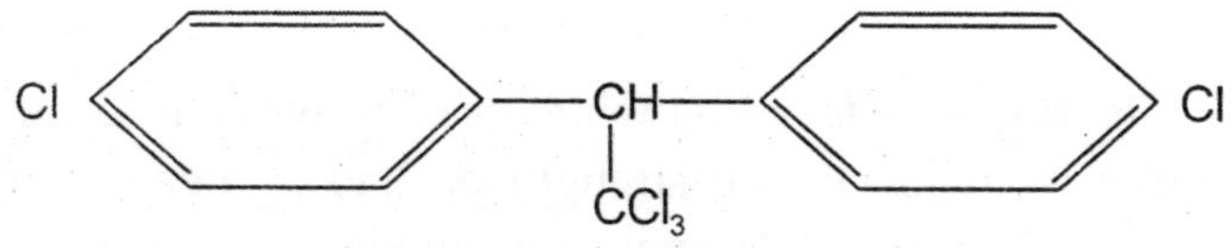

The technical DDT has a setting point of 89% and should contain 9.5 to 11% hydrolysable chlorine and less than 1% volatile and alcohol insoluble material. DDT is almost completely insoluble in water (0.00001%), fairly soluble in most organic solvents and highly soluble in aromatic hydrocarbons such as xylene. In natural oils, e.g., linseed, pea-nut, cotton seed oil, the solubility is 10 to 12%.

It is readily dehydrochlorinated in alkaline solutions to form a non-insecticidal product namely dichloro ethylene. It is generally stable under most conditions and has a very low vapour pressure. DDT is a contact and stomach poison has long residual action. It affects the sensory organs and nervous system and causes violent agitation which is followed by paralysis and death. It is slow acting insecticide relatively nontoxic to mammals but in oil solution it is absorbed by the skin. When animals are fed with treated foliage, accumulation of the DDT in animal fat and its secretion in the fat (butter) of lactating animal has been observed. DDT is relatively safe for use on large number of crops but exhibits pytotoxicity on sensitive crops like cucurbits. It is also found to stimulate plant growth in some crops like potato, tomato, brinjal, cabbage etc. at lower concentrations. DDT is effective against a wide range of insect pests although some of them (singhara beetle, cutworm, sugarcane pyrilla and mango jassids etc.) have developed resistance to it. Since DDT kills natural enemies, hence some insects such as aphids, scale insects mealy bugs, grasshoppers and mites develop in more number after its application. The oral and dermal LD 50 for rat are 300-500 and 2500 mg/kg respectively. It can be used in combination with many insecticides except lime sulphur or any strong alkaline material in which it decomposes rapidly DDT is available in the following form :

Dusting powder : It is white or dirty white in colour and mainly available in the strength of 2 to 10 per cent of the toxicant. The dust is prepared by mixing the chemical with inert material like talc, gypsum and pyrophillite, etc. It is applied generally at the rate of 25 to 30 kg per hectare against a variety of crop pests, *viz.*, swarming caterpillar, army worm, rice-case worm, cut worms, potato beetles, cotton leaf roller etc.

Water-dispersible powder or wettable powder : It is used for preparing suspensions in water and is available in a white powder containing 50% of actual DDT. It is used as 0.15 to 0.32% spray for the control of most agricultural pests.

Emulsifiable concentrate : DDT is available as 25% emulsifiable concentrate and dark brown in colour, It is used at 0.15 to 0.32% concentration for the control of pests. Due to larger residual action, emulsions preferred to the water-despersible powder.

Several DDT analogues which have insecticidal properties are also used to control insects. Among them TDE or DDD, perthane, methoxychlor, dilan, dicofal and DMC are most common.

Monocyclic Structure

These compounds contain only one non-aeromatic hydrocarbon ring structure.

BHC-BHC is 1, 2, 3, 4, 5, 6-Hexachlorocyclohexane which has the empirical formula $C_6H_6Cl_6$. Its common name benzene hexachloride is incorrect from a chemical stand point as HCH could be confused with hexachloro benzene. This compound was first prepared by Michael Faraday in 1825. Its insecticidal properties were discovered simultaneously in England and France in about 1942. Its structural formula is as follows :

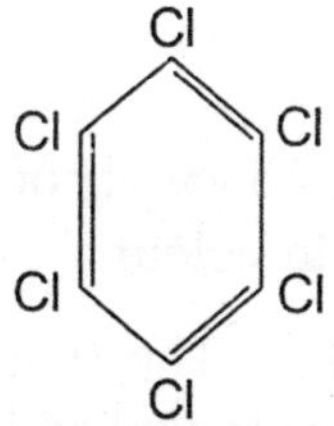

It is produced by the chlorination of benzene in the presence of sunlight. The technical product is greyish or brownish amorphous solid and has a strong musty odour. Among the 13 possible isomers found in varying proportions in BHC, the gamma isomer is insecticidically most active and the percentage of following 6 isomers in BHC is as follows :

Isomers	*Melting point in °C*	*Percentage in BHC*
Alfa	157	55-77
Beta	309	5-14
Gamma	112	10-18
Delta	138	6-8
Epsilon	219	3-4
Eta	90-91	Unknown

The 6 isomers differ widely in their solubilities in the common organic solvents, in their degree of volatility and in their stability to alkali. BHC is stable compound but is broken down by alkali. It acts as a stomach and contact poison and also to some extent as a fumigant. BHC is more toxic to insects than DDT but less poisonous to man and domesticated animals than DDT. BHC poisoning increases respiration rate with the symptoms of tremors, ataxia, convulsions and prostraction. BHC is more volatile than DDT and therefore, does not have a long residual toxicity. In closed places it is effective about a week and in field 3-4 weeks. It is available in the following formulations.

Dust : The dusting powder of BHC contains five to ten percent of the active ingredient and is commonly used @ 25 to 30 kg/hect to control a variety of crop pests. The dust is light grey or light brown in colour.

Wettable powder : The water-dispersible powder is available in 50% concentration and light brown in colour. It is used in spray in the concentration of 0.15% to 0.5% against aphids, jassids, white flies, leaf rollers, leaf cutting beetles and thrips etc.

Emulsifiable concentrate : Its emulsion is dark brown to light brown in colour and is available as 20% emulsifiable concentrate. It is used in the same concentration as wettable powder against the same pests.

BHC can be used with other insecticides and fungicides except those having strong alkaline materials, *e.g.*, Bordeaux mixture, lime or lime-sulphur.

Lindane

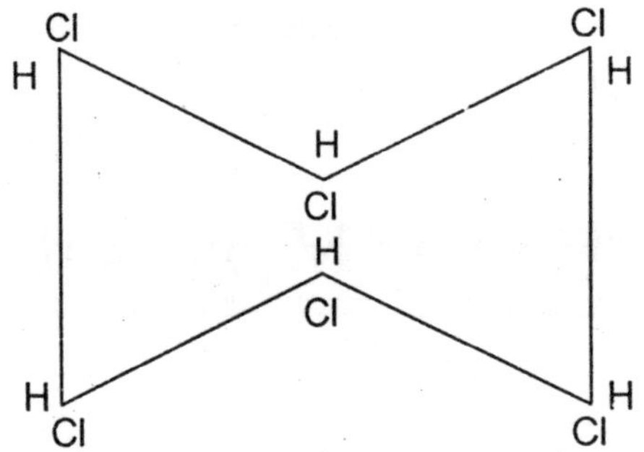

It is also known as gamma BHC as it contains 99% gamma isomer. The name lindane was proposed in 1949 after Vander Linden, a German Chemist who isolated this isomer in 1912. It is colourless crystals with a slight aeromatic odour. The lindane is practically insoluble in water (10 parts per million), slightly soluble in common paraphenic hydrocarbons and soluble in common organic solvents. It is a contact and stomach poison and has some fumigant properties. It is unstable in the presence of alkali and does not accumulate in the fat of animals. It can be used for controlling a wide range of vegetable pests on which BHC can not be used. The lindane is formulated as 20% emulsifiable concentrate, 10% granule and 1 to 3 % dust. Like BHC, it is also less poisonous to man and domesticated animals than DDT. Its oral and dermal LD 50 values to rat are 200 mg/kg and 500-1,000 mg/kg respectively.

Bicyclic Structure

These insecticides contain 2 non-aeromatic hydrocarbon rings fuse together so as to have 2 non-adjacent carbon atoms common to both rings. These are produced by the chlorination of the naturally occurring terpenes hence also called as chlorinated terpenes. It includes toxaphene and strobane.

Toxaphene ($C_{10}H_{10}Cl_8$)

This insecticide is a chlorinated camphene containing 67 to 69% chlorine. A technical product is a yellowish waxy solid melting point 65 to 90°C and is a mixture of isomers. It is unstable in the presence of alkali and upon prolonged exposure to sunlight. It is insoluble in water, readily soluble in paraphenic hydrocarbons and most soluble in organic solvents such as benzene, xylene, ethylene dichloride and carbon tetra chloride. Its structural formula is as follows :

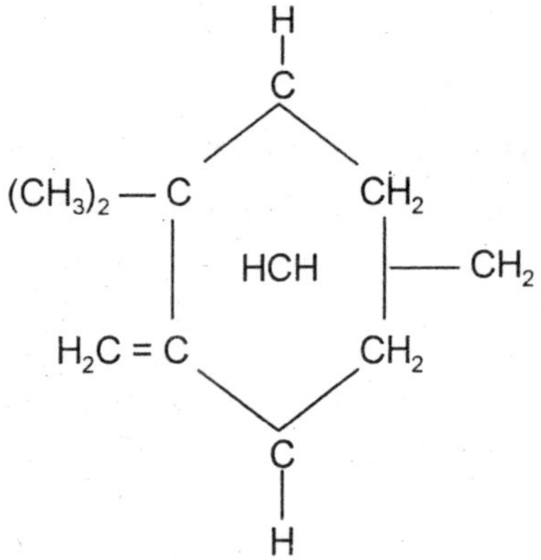

Toxaphene primarily acts as a stomach poison but some times is also used as contact poison. In comparison to BHC and DDT, it is slightly slow in action and has a lesser residual toxicity. It is non-toxic to honey bees but to human beings, it is four times more toxic than DDT. Toxaphene shows pytotoxicity to cucurbits and is also poisonous to fish. It is formulated as follows :

1. Dusting powder—It is available 5, 10 and 20%.
2. Water dispersible powder—25 to 40%.
3. Emulsifiable concentrate —Available as 25%.

Strobane ($C_{10}H_{11}Cl_7$)

It is prepared by chlorinating a mixture of camphene and pinene and containing 66% chlorine. The technical material is straw coloured liquid. It is practically insoluble in water but soluble in most organic solvents. It is not more stable to alkali. It has strong insecticidal and some acaricidal properties. Its oral and dermal LD 50 values, respectively, are 200 to 250 mg/kg for rat and 500 mg/kg for rabbit.

Polycyclic Structure

These compounds contain 3 or 4 non-aeromatic hydrocarbon rings with atleast two points of fusion. The insecticides of this group have been derived from a hexa chlorocyclo pentadiene and some times also called as the cyclodiene insecticides which also include the some insecticides of the next group (Heterocyclic structure). It includes following three insecticides.

Aldrin ($C_{12}H_8Cl_6$)

It is chemically 1, 2, 3, 4, 10, 10- hexachloro- 1, 4, 4a, 5, 8, 8a- hexahydro- 1, 4, -endo-exo-5, 8- dimethanonaphthalene. Aldrin is white crystalline compound with melting point 104°C, and containing 78% active aldrin. It is soluble in organic solvents such as acetone, amyl acetate, benzene, carbon tetrachloride, ethyl alcohol and xylene etc. Its structural formula is as follows:

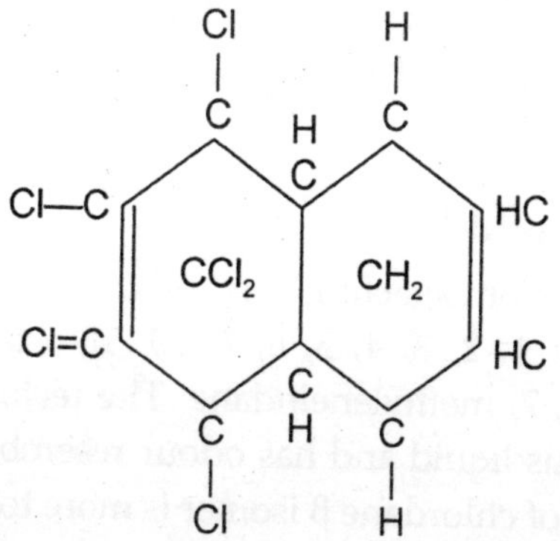

Aldrin is stable in strong alkali and is compatible with all commonly used insecticides. It is non phytotoxic to plant under most conditions of uses and free from smell. Aldrin is more effective insecticide than other chlorinated insecticides and has proved particularly valuable for the control of soil borne pests (cutworms, white grubs, termites and grasshoppers etc.) because of its prolonged residual action. It mainly acts as a contact or stomach poison and is more toxic than DDT. Its oral and dermal LD 50 values for rat are 40 to 50 and 200 mg/kg, respectively. Crops meant for human or cattle consumption should not be sprayed or dusted with aldrin three weeks prior to harvesting. It is formulated as under :

(i) *Dust* : Usually 5% dusting powder is available and used at the rate of 25-30 kg/hect.

(ii) *Wettable powder* : Water dispersible powder of 40% is formulated which is applied at a concentration of 0.15 to 0.3%.

(iii) *Emulsifiable concentrate* : It is available as 30% and used in the same concentration as wettable powder.

Note : Aldrin 5% dust is generally used for the control of cut-worms attacking potato crop and is applied in the soil at the time of sowing. Recently, the experiments conducted at Indian Agriculture Research Institute, New Delhi, revealed that the residues of aldrin were present in the tubers 10 times more than the permissible limit, besides it was also present in the form of dieldrin. Therefore, the aldrin should not be used in tuber crops.

Chlordane ($C_{10}H_6Cl_8$)

Chlordane or Octachlor was discovered by Julious Hyman in 1945. It is 1, 2, 3, 4, 5, 6, 7, 10, 10-octachloro-4, 7, 8, 9-tetrahydro-4, 7, methyleneindane. The technical material is a brown viscous liquid and has odour resembling cedar. Of the two isomers of chlordane β isomer is more toxic than α isomer.

The technical material is unstable in the presence of alkaline material liberating hydrogen chloride and forming non-toxic product. Its specific gravity is 1.61. It is insoluble in water but highly soluble in common solvents. Chlordane is slowly volatile and, therefore, has a slight fumigant action besides being a contact and stomach poison. Its structural formula is as follows :

Chlordane should not be used on leafy vegetables or on root crops since it is likely to taint them. At higher concentrations it is toxic to cucurbits. It is also equally toxic to warm blooded animals as DDT and hazardous to fish. Its oral and dermal LD 50 values for rat are 283 and 1600 mg/kg respectively. It is formulated as under :

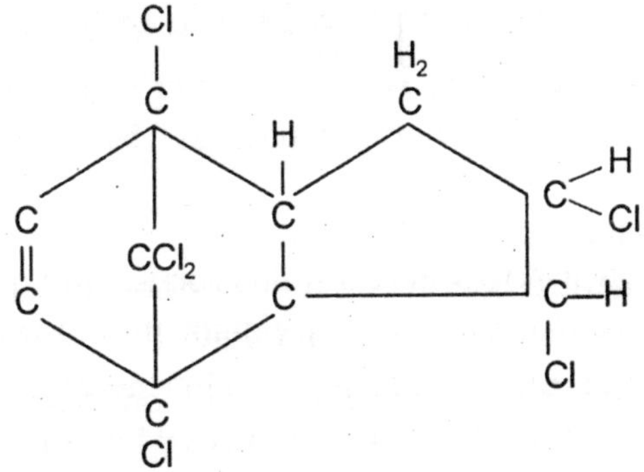

(i) *Dust* : Dusting powder is generally available as 5% and is used at the rate of 25 to 30 kg per hectare to control the grasshoppers and termites.

(ii) *Emulsifiable concentrate* : It is formulated as 20% and used in the strength of 0.15 to 0.30%.

(iii) *Wettable powder* : It is not in much use, previously it was formulated as 40 to 50% and used in the same strength as 'emulsifiable concentrate.

Heptachlor ($C_{10}H_5Cl_7$)

Heptachlor is 1, 4, 5, 6, 7, 10,10 heptachloro-4, 7, 8,

9-tetrahydro-4, 7-methylenein-dane. The technical product contains about 67% heptachlor and is waxy solid. The purified material is white crystalline solid with melting point 95-96°C. It is practically insoluble in water but variously soluble in acetone, benzene, CCl_4, ethyl alcohol, deodorised kerosine and xylene etc. It is stable to air, light, heat, moisture and mild alkaline conditions. Its structural formula is as follows :

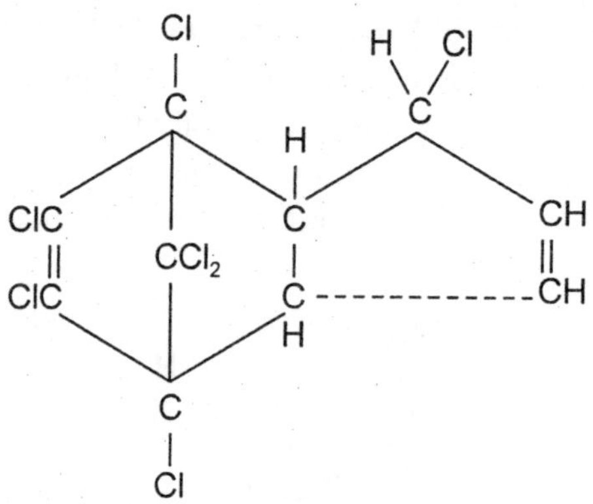

Heptachlor has strong insecticidal properties with possibly some fumigant action. It is a nerve poison and about 2 times more toxic to mammals than chlordane. Its oral and dermal LD 50 values for rat are 40 and 200-250 mg/kg respectively. Heptachlor residues are slowly converted to heptachlor epoxide in and on plants and in animal tissues and has the same mammalian toxicity. It is formulated as under :

(i) *Dust* : The dusting powder is available in the strength of 3-5% and used @ 25 to 30 kg per hectare to control the soil inhabiting pests as cutworms, wireworms, white grubs and termites etc.

(ii) *Emulsifiable concentrate* : It is formulated in the strength of 20% and used as a spray in the concentrations ranging from 0.10 to 0.30%.

(iii) *Wettable powder* : Available as 25% and used in the same concentration as emulsifiable concentrate.

Heterocyclic Structure

These compound consist of 1 or more rings as in the polycyclic structure but atleast one of the atoms of the ring is not a carbon atom, *i.e.*, it is usually N, O or S. It includes following insecticides :

Dieldrin ($Cl_2H_8Cl_6O$)

This is the epoxide of Aldrin and chemically it is 1, 2, 3, 4, 10, 10-hexa chloro-6, 7-epoxy-14a, 5, 6, 7, 8a-octahydro-l, 4-endo, exo-5, 8 dimethano naphthalene. The technical material is containing about 76% of active ingredient. The technical material is light tan flaky solid and when purified it exhists as white odourless crystals melting at 176°C. It is insoluble in water but moderately soluble in organic solvents. Its structural formula is as follows :

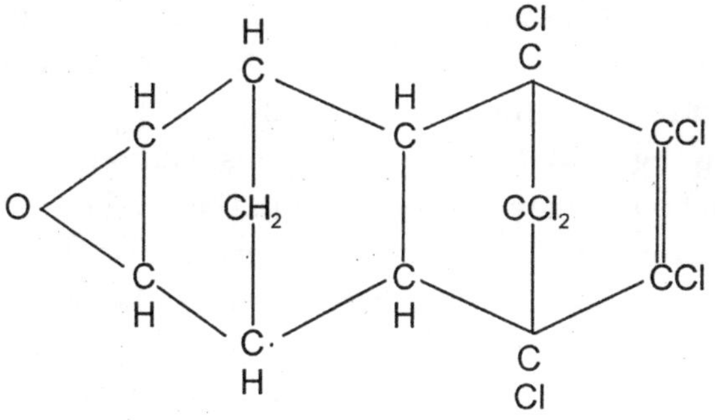

Dieldrin is very stable material and is not affected by alkali, light and mild acids. It has long residual action and is compatible with most of the insecticides and fungicides. It acts as a stomach and contact poisons and 10 times more toxic than Aldrin. Its oral and dermal LD 50 values to rat are 40 and 100 mg/kg, respectively. It is available in the following formulations and used to control the chilli thrips, termites, ants, grasshoppers, cutworms etc.

(i) *Dust* : 1.5% dusting powder is available which is used @ 25 to 30 kg/hect.

(ii) *Wettable powder* : It is formulated in the strength of 20 to 50% and used in the concentration of 0.05 to 0.12%.

(iii) *Emulsifiable concentrate* : It is available in the strength of 20% and used in the same concentrations as above.

Endrin ($C_6H_8Cl_6O$)

It is 1, 2, 3, 4, 10, 10-hexa chloro 6, 7-epoxy-l, 4, 4a, 5, 6, 6, 8, 8a-Octahydro-exo-l, 4-endo, endo-5, 8-dimethamonaphthalene or the endo, endo isomer of dieldrin. The technical endrin is less stable than dieldrin and its formulations is stabilized by the addition of small amount of hexa methylene tetramine. It is insoluble in water, slightly in alcohols and petroleum oils and moderately soluble in acetone and benzene. Its structural formula is similar to dieldrin.

Endrin has a wide range of spectrum and high residual toxicity. It acts as a contact and stomach poison and also has a fumigant action. It is the most toxic of the hydrocarbon insecticides, its acute oral and dermal LD 50 values to rat are 3-6 mg/kg and 60-120 mg/kg respectively. Insecticidal formulations containing endrin should be handled with extreme care because of its great toxicity to mammals. The use in food and forage crops nearing harvest should be avoided. It has gained great popularity for the control of sugarcane and paddy borers, spotted boll worms, cotton jassids, thrips and sugarcane leaf hoppers etc. It is available in the following formulations:

(i) *Dust* : Available as 1% and used at the rate of 25 kg/ hect.

(ii) *Emulsifiable concentrate* : It is formulated as 20% emulsifiable concentrate which is used in the concentrations of 0.01 to 0.03% for the control of a wide range of pests in India.

(iii) *Granule* : Endrin 2% granule formulation is available which has been to be effective for the control of Lepidopterous borers infesting sugarcane, sorghum and rice etc.

Note : The government of India has imposed restriction since 1st April, 1977 in the use of endrin due to its high mammalian toxicity.

Thiodan *($C_9H_6Cl_6O_2S$)*

This insecticide commonly known as endosulfan and contains both S and O, as heterocyclic atoms. Chemically it is 6, 7, 8, 9, 10, 10-hexachloro- 1, 5, 5a, 6, 9, 9a-hexahydro-6, 9-methano-2, 4, 3-benzo dioxathiepin 3-oxide. The technical material is brownish crystalline solid melting from 70-100°C. It is both a chlorinated hydrocarbon and an organic sulphite. The structural formula is as follows :

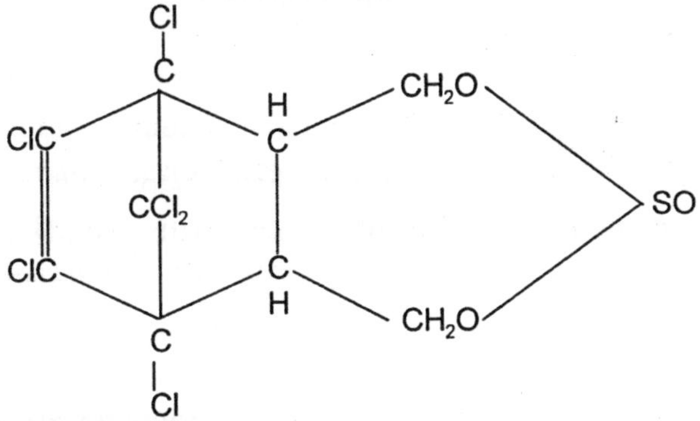

It is practically insoluble in water but moderately soluble in common organic solvents. It acts as contact and stomach poison and also has slight fumigant action. Thiodan is general purpose insecticide effective against many different types of insect pests. It is not harmful to bees and natural enemies of pests. Its acute oral and dermal LD 50 values of rat are 110 and 74 to 130 mg/kg respectively. It is specially effective against borers, aphids, leaf hoppers and other Homopterous pests. It is available in the following formulations :

(i) *Dust* : 4% dust generally available which is used @ 25-30 kg/hect. It can be safely used on cucurbits to control the red pumpkin and epilachna beetles.

(ii) *Emulsifiable concentrate* : It is available in the strength of 35% and used in the concentration of 0.05 to 0.07%.

(iii) *Granule* : 4% granules are available and used for the control of soil inhabiting pests and stem borers of sorghum and sugarcane.

Organo-phosphorus insecticides : The insecticides of this group contain one or more atoms of phosphorus bonded either directly to the carbon atoms or indirectly through N_2, O_2 or S. Schradar and his associates just prior to IInd world war discovered the insecticidal potency of this chemical group in Germany. The insecticides of this group mainly act as contact and stomach poisons, and some are systemic, some slowly give out vapours which have fumigant action. All the organophosphorus insecticides appear to have a common mode of action.

They are essentially irreversibly inhibitors of cholinestrase enzyme of the neuro-muscular system. They have the decided advantage of being detoxify very rapidly to animal tissues and eliminated rather than these stored in fatty tissues as most of the chlorinated hydrocarbon insecticides.

The insecticides of this group are classified on the basis of phosphorus atoms found in the combination as phosphatase (TEPP, DDVP, chlorfenvinphos, tetrachlorvinphos, phos-phamidorr, monocrotophos, dicrotophos and mevinphos), phosphonates (trichlorphon), phosphorothionates (parathion, fenitrothion, fenthion fensulfothion, chlorpyriphos, thionazin, diazinon, methyl-demeton, demeton and quinalphos), phosphorothioiates + (oxydemeton-methyl) phosphorothiolothionates (ethion, phorate, thiometon, disulfoton, dimethoate, formothion malathion, phenthoate, carbophenothion and phosalone), phosphonothionates (EPN and leptophos), Phosphoramidothioate (methamidophos), phosphoramidate (schradan, mephosfolan and phenamiphos).

On the basis of their mode of action these insecticides may be classified into two groups.

Contact Insecticides

It includes following chemicals :

1. HETP (*Hera ethyl tetra phosphate)*
2. Sulphotate
3. Parathion
4. Ethyl parathion
5. Methyl parathion
6. Malathion
7. Diazinon
8. Guthion
9. Trithion
10. Ethion
11. Dipterex
12. DDVP
13. Dibrom
14. Dursban
15. Zolone
16. Oftanol
17. Volaton
18. Hostathion
19. Tamaron
20. Elsan

Systemic Insecticides

These are those chemicals which are absorbed by the plant system and translocated in the whole parts of the plants. The some of the important insecticides are as follows :

1. Shradan
2. Dimefox
3. Demeton
4. Methyl-o-demeton
5. Rogor
6. Phosdrin
7. Dimecron
8. Phorate
9. Disyston
10. Ekatin
11. Dasanit
12. Azodrin
13. Cytrolane
14. Formothion
15. Phosvel

The important and commonly used insecticides are described below :

Malathion ($C_{10}H_{19}O_6PS_2$)

The technical material is 95% to 98% pure with an unpleasant odour. The purified compound is brownish liquid and chemically it is 0, 0-dimethyls (1, 2-dicarbe thoxyethyl) Phosphorodithioate. It is soluble in water to only 0.0145%, slightly soluble in mineral oils and soluble in most organic solvents.

It is unstable in alkaline medium. Its structural formula is as follows :

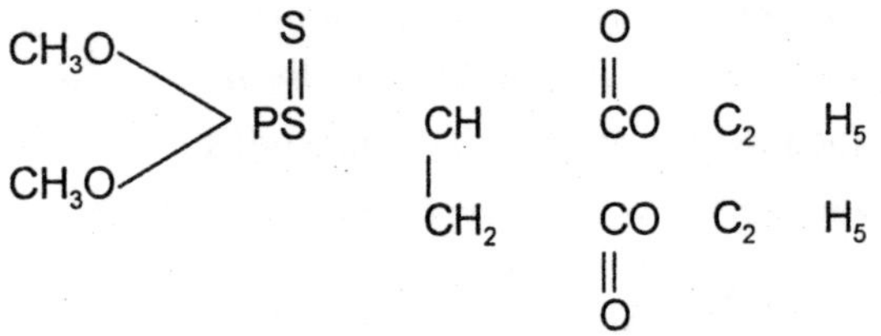

Malathion is mostly used as contact insecticide though it acts as a stomach poison and exhibits some fumigant action also. It is used for the control of a wide variety of pests of crops such as aphids, hoppers, psyllids, bugs, caterpillars, floora-beetles and mites etc. The malathion is the safest insecticide available so far, its acute oral and dermal LD 50 to rat being 1400 to 1900 and 4000 mg/kg respectively. At the recommended dosages it can be applied to most edible crops within one to three days of harvest without danger of excess harmful residues. It is formulated as dusts, wettable powders and emulsifiable concentrates and is being marketed as follows :

1. Dusting powder-5 per cent used 25 to 30 kg/hect.
2. Water dispersible powder-25 per cent and used in concentration of 0.025 to 0.03%.
3. Emulsifiable concentrate-50 per cent used in same concentration as above.
4. TJLV preparation : 96 per cent used in aerial spraying.

Diazinon ($C_{12}H_{21}N_2O_3PS$)

Chemically it is diethyl *o*-isopropyl-4 methyl-6-pyrimidinyl phosphorothionate and also known as Basudin. Diazinon is a colourless liquid with a slight characteristic odour and technical product is 90 per cent pure. It is slightly soluble in water (0.004%) and soluble in most of the organic solvents. Its structural formula is as follows :

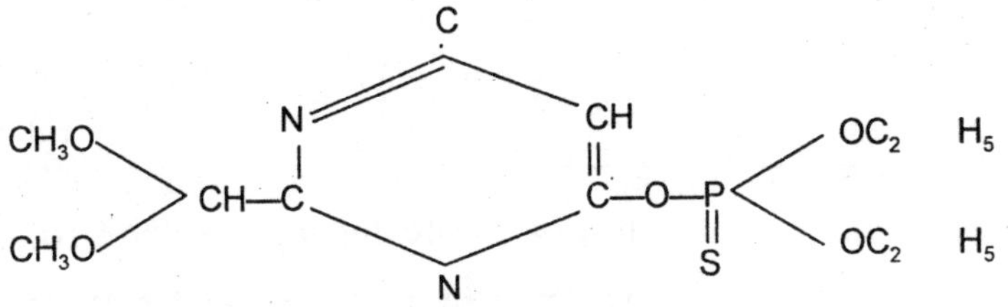

Diazinon is a strong contact and stomach poison and has fumigant action and penetrating quality. Its acute oral and dermal LD 50 for rat are 150 to 220 and 500 to 2000 mg/kg respectively. It is used to control a variety of agricultural pests, *viz.*, lepidopterous larvae, bugs, thrips, beetles and mites etc. It has proved to be an outstanding insecticide for the control of most sucking insects, borers and leaf caterpillars. It is formulated as follows :

1. *Dust* : 2 to 5 per cent dusting powder and used @ 25 kg/hect.
2. *Emulsifiable concentrate* : 25 per cent used in the concentration of 0.025 to 0.03 percent.
3. *Granule* : 5 to 10 per cent and used @ 25 kg/hect to control the soil inhabiting pests.

Trichlorfon ($C_4H_8CI_3O_4P$)

Chemically it is 0, 0 dimethyl 2, 2, 2-trichloro hydroxyethyl phosphonate. Purified chemical is a white crystalline solid, melting point 83-84°C. It is moderately soluble in water (15 per cent) and freely soluble in most organic solvents. Its structural formula is given below :

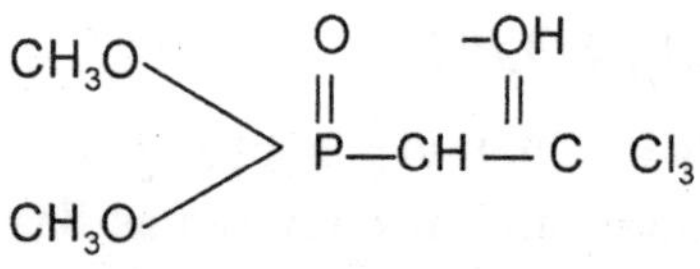

Trichlorfon is a contact and stomach poison with some fumigant action and has some penetrating effect also. It is a quick acting chemical effective against Lepidoptera, Diptera and Hemiptera. Under natural conditions it is readily dehydrochlorinated to dichlorvos. It has low mammalian toxicity but toxic to insects and effective up to 10 days of application. Its acute oral and dermal LD 50 values for rat are 560 and 72,800 mg/kg. Trichlorfon is also used in baits to control the fruit flies. It is generally available as 5% dust and 80 per cent wettable powder. Its trade names are dipterex; dylox and tugon.

Dichlorvos or DDVP ($C_4H_7O_4Cl_2P$)

Chemically it is 2, 2 dichlorovinyl dimethyl phosphate. It is one of the products resulting from the alkaline treatment of trichlorfon. DDVP is thin colourless liquid with pleasant smell, boiling point 120°C. It is insoluble in water but readily soluble in most organic solvents. Its structural formula is as follows :

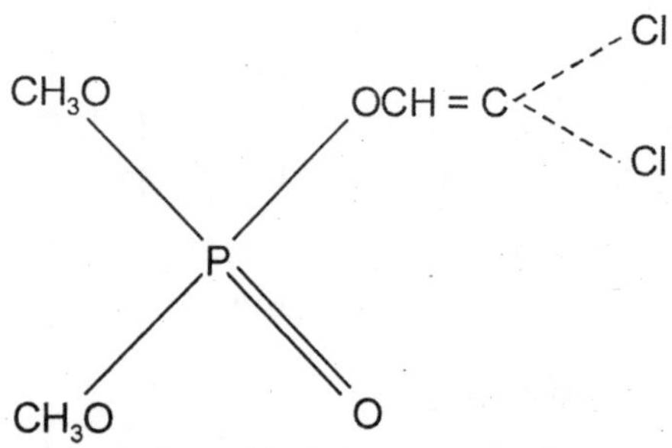

Dichlorvos gets hydrolysed slowly in neutral and acidic media and rapidly in alkaline medium. It is contact and stomach poisons with some fumigant action and has good penetration properties.

It brings about rapid knock down of the insects. It is highly volatile compound, therefore, can be used on all crops just before harvest. DDVP is very effective in baits and in aerosol formulations for the rapid knock down of flies, mosquitoes and moths etc. Its acute oral and dermal LD 50 values for rat are 25-30 and 75 to 900 mg/kg. It is available in the form of 78 EC with the trade names Nuvan and Vapona etc.

Carbophenothion

Chemically it is *o*, *o*-dimethyl-S para chlorophenyl thiomethyl phosphorodithioate. It is light amber colour liquid insoluble in water and soluble in common organic solvents including the paraphenic hydrocarbons. The pure chemical is a colourless, odourless oil. It hydrolyses very slowly in water suspension. Its structural formula is as follows :

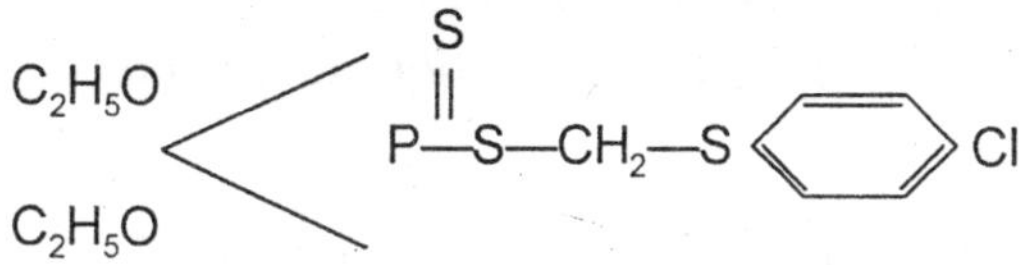

It is a general purpose insecticide with long residual action in the field and effective against a wide range of insects and mites, particularly aphids, thrips, scale insects, leaf hoppers etc. Its acute oral and dermal LD 50 values to rat are 7-30 and 800 mg/kg respectively. It is formulated as 15 per cent EC and 25 per cent W.P. and sold in the names of Trithion and Garrathion.

Ethion ($C_2H_{22}O_4P_2S_4$)

Chemically it is 0, 0, 0, 0-tetraethyl S, S-methylene bis phosphorodithioate. Ethion is a colourless liquid insoluble in water but soluble in most organic solvents. It hydrolyses under both acidic and alkaline conditions.

Its structural formula is as follows :

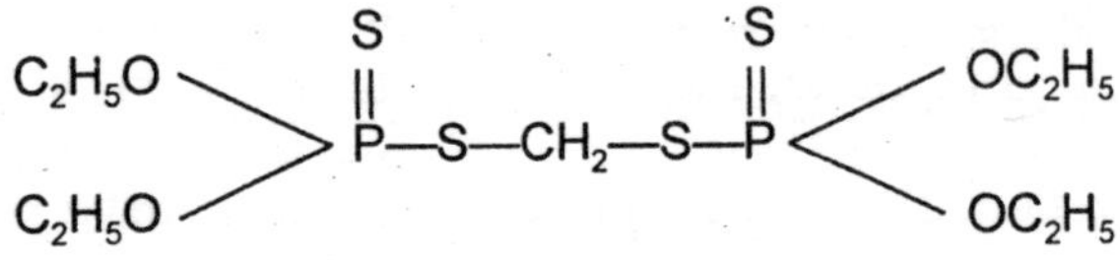

It is useful for the control of many insects, *viz.*, aphids, scale insects, leaf hoppers, leaf miners, thrips, bugs and mites. Effectiveness of ethion for control of scale insects and mites appear to be more by formulating it in petroleum oil. It is generally formulated as 50% EC, 25 to 50% W.P. and 2% solution in spray oil.

Chlorpyrifos ($C_9H_{11}CI_{13}NO_3PS$)

Chemically it is *o*, *o*-diethyl *o*- (3, 5, 6-trichloro-l-pyridyl) phosphorothioate. Chlorpyrifos is white crystalline substance with mild mercaptan odour. It is practically insoluble in water but readily soluble in most organic solvents. It is relatively stable to hydrolysis and oxidation. Its structural formula is as follows :

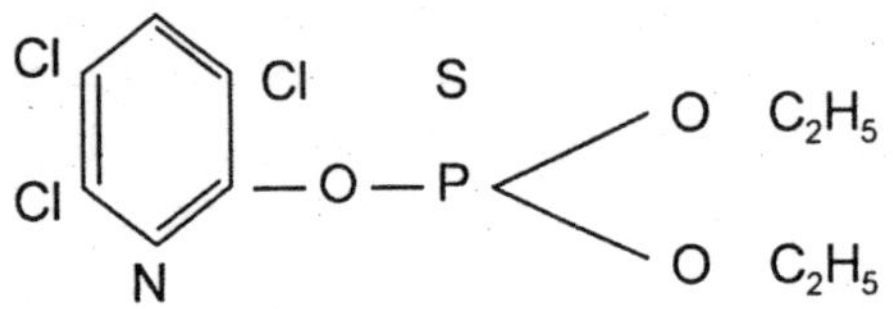

It is a contact and stomach poison with vapour action also and effective against many pests with particular reference to rice pests, borers, aphids, mosquitoes, soil inhabiting insects and some mites. It is formulated as 20 EC and available in the name of Dursban. Its acute oral and dermal LD 50 values to rat and rabit are 135 mg/kg and 1000 to 2000 mg/kg respectively.

Quinalphos ($C_{12}H_{15}O_3H_2PS$)

Chemically it is *o*, *o*-diethyl-o (quinoxalinyl, (2) thionophosphate. Quinalphos is an odourless white crystalline solid, melting point 31-32°C. It is practically insoluble in water but readily soluble in organic solvents. It is hydrolysed in a alkaline medium. Its structural formula is as follows :

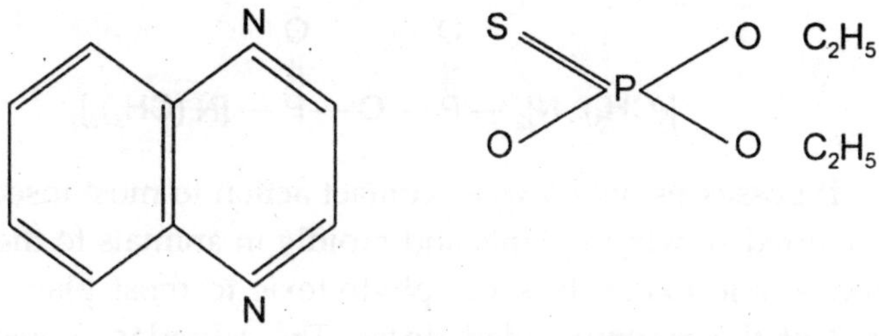

It is a strong contact and stomach poison and effective against a wide spectrum of insect pests of crops. Quinalphos is also used as effective acaricide. It is formulated as 25 per cent EC and 5 percent granule and available in the name of Ekalux. Its acute oral LD 50 for male rat is 137 mg/kg and dermal 1400 mg/kg.

Systemic Insecticides

Those insecticides which are translocated to all the parts of the plants when applied on the roots, stems, leaves or seeds, are known as systemic insecticides. The important and commonly used insecticides of this group are described here. These insecticides have following merits :

1. Equal coverage is possible.
2. Protecting new plant growth.
3. Having less harmful effects on beneficial, predatory and pollinating insects.
4. Increasing the residual effect as less affected by weathering.

Scharadan

It is octamethyl pyrophosphoramide and was among the 1st systemic insecticides. The technical material is brown liquid and contains 40 per cent of schradan boiling point 154°C. It is soluble in water and also remarkable stable to alkaline hydrolysis. Its structural formula is as follows :

$$[(CH_3)_2N]_2 - \overset{\overset{\displaystyle O}{\|}}{P} - O - \overset{\overset{\displaystyle O}{\|}}{P} - [N(CH_3)_2]_2$$

It possesses only a week contact action to most insect and converted slowly in plants and rapidly in animals to the toxic monoamide oxide. It is not phyto-toxic to most plant when used at the recommended doses. The schradan is specially effective against aphids and mites. Its acute oral and dermal LD 50 values to rat are 5 and 50-100 mg/kg respectively. It is also known as Pestox III. Systam and OMPA.

Demeton

It is mixture of 2 parts of 0, 0-diethyl 0-2 (ethylthio) ethyl phosphorothionate and I part of demeton-s-ie, diethyl 2-(ethyl thio) ethyl phosphorothionate. The technical material is yellowish liquid, soluble in organic solvents and unstable in alkaline solution. Both the isomers are only slightly soluble in water. Its structural formula is as follows :

$$\text{2 parts } (C_2H_5O)_2\overset{\overset{\displaystyle S}{\|}}{P}OCH_3CH_2SC_2H_5$$

$$\text{1 part } (C_2H_5O)_2\overset{\overset{\displaystyle O}{\|}}{P}SCH_2CH_2SC_2H_5$$

Demeton provides a long systemic action rapidly observed by roots, stems or foliage. It is particular effective as a spray for the control of many species of aphids and red spider mites and has no marked phyto-toxic effect. Its acute oral and dermal LD 50 values for rat are 3-5 and 200 mg/kg, respectively. It is formulated as 25 Ec and sold in the trade name of systox.

Demeton-o-methyl

It is 0, 0-dimethyl s-2-(ethyl-thio)-ethyl phosphorothionate. It has very similar property in behaviour as demeton but less toxic than the former. Its acute oral and dermal LD 50 values for rat are 50 to 75 and 300 to 450 mg/kg respectively. Its structural formula is as follows :

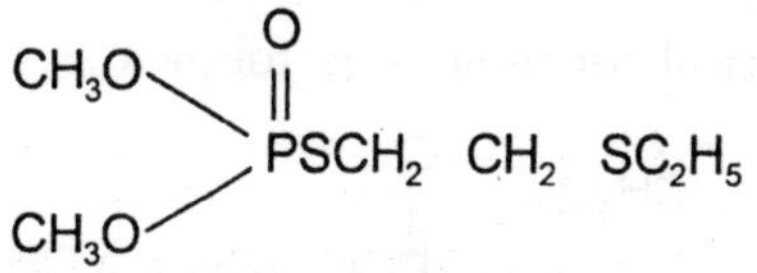

It is effective against sucking insect pests and mites of crops. It is formulated as 25 per cent EC and available in the name of Metasystox.

Dimethoate ($C_5H_{12}O_3PS_2N$)

Chemically it is *o*, *o*-dimethyl S (N-methyl carbomoylmethyl) Phosphorodithioate. Dimethoate is a white crystalline solid, melting point 121.8 to 123.6°F and technical material contains 94 to 96 per cent chemical. It is soluble in organic solvents and also in water to the extent of about 7% and unstable in alkaline media. Its structural formula is as follows :

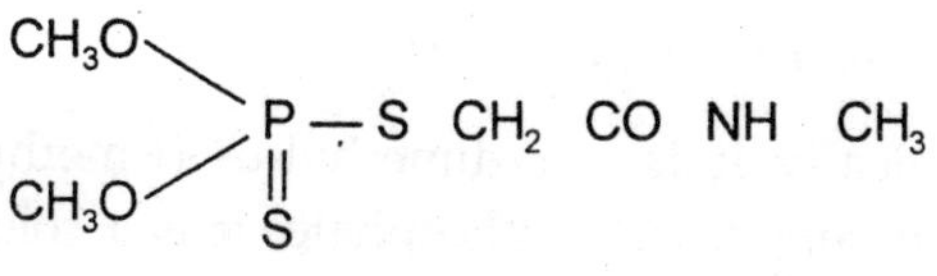

It is a systemic and contact poison effective against sucking insect pests and mites of a wide variety of crops and fruit trees. It is highly toxic to insects but less toxic to mammals, its acute oral and dermal LD 50 values for rat are 200 to 300 and 700 to 1150 mg/kg respectively. It is generally formulated as 30 per cent EC and sold in the name of Rogor.

Mevinphos

Chemically it is dimethyl 2-carbomethoxy-l-methylvenyl phosphate. The technical material is colourless liquid, melting point 106-107°C miscible in water and soluble in most of the organic solvents. The technical product is composed of about 2 parts of *trans*-isomer and I part of *cis*-isomer. The *trans*-isomer is about 10 times as active as *cis*-isomer.

Its structural formula is as follows :

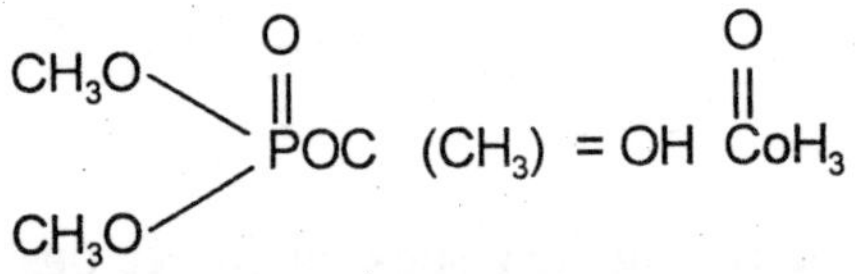

It is a systemic insecticide and has powerful contact and fumigant action. It is rapidly broken down in plants, therefore, its residues are very short lived in the field and is specially useful to control the pests on food crops close to harvest. It is highly poisonous by injection, absorbtion through skin and inhalation.

Areas inhabited by wild life and fish should not be contaminated with this. Its acute oral and dermal LD 50 values to rat are 3 to 5 and 90 mg/kg respectively. It is formulated as 25 per cent EC and 10 per cent W.P. and sold in the name of phosdrin.

Phosphamidon ($C_{10}H_{19}O_5NCIP$)

Chemically it is *o, o*-dimethyl-o- (I-methyl-2-chloro-2-diethyl-carbamoyl-vinyl) phosphate. It is a colourless liquid, boiling point 160°C with faint pleasant odour. Technical grade is dark brown and commercial product is bright violet due to addition of a dye. It is miscible with water and all organic solvents except saturated hydrocarbons in which it is soluble only to a limited extent.

Its structural formula is as follows:

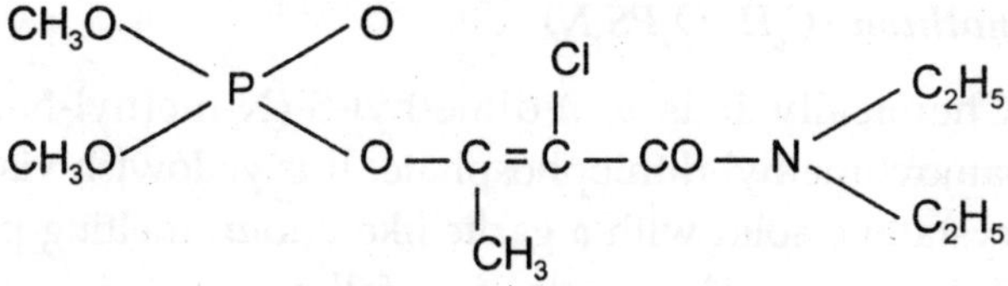

Phosphamidon is a systemic insecticide and acts as a stomach poison and has a relatively low contact action. It is rapidly absorbed by the plant surface and is quickly decomposed in the plant. Metabolism of phosphamidon gives rise to desethyl phosphamidon which is roughly as toxic as the present compound. It is effective against many sucking, chewing and mining insects and certain mites. Its acute oral and dermal LD 50 values for rat are 15 and 125 mg/kg respectively. It is formulated as 85% and available in the trade name of Dimecron.

Monocrotophos ($C_7H_{14}O_5NP$)

Chemically it is dimethyl *cis*-l-methyl-2-methyl carbamoyvinyl phosphate. It is made up of colourless crystals with mild ester smell, boiling point 125°C. Monocrotophos is miscible with water and all organic solvents. It is rapidly hydrolysed in alkaline solutions than in acids. Its structural formula is as follows :

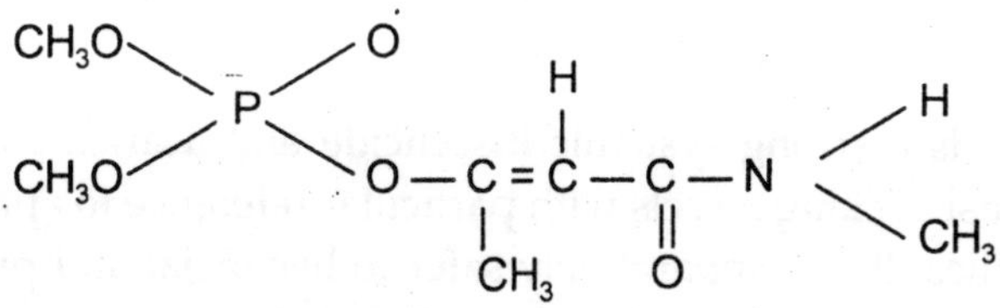

Monocrotophos is a systemic insecticide but also acts as a stomach and contact poison. It is a hazard to wild life if the food is contaminated with residues above 0.5 ppm. Its acute oral LD 50 value for rat is 20 mg/kg. It is effective against a wide variety of insect pests, *viz.*, thrips, aphids, leaf miners, cotton boll worm, white flies and some mites. It is generally available as 40% SC and sold in the name of Azodrin and Nuvacron.

Formothion ($C_6H_{12}O_4PS_2N$)

Chemically it is *o, o*-dimethyl-S-(N-methyl-N-formyl) carbamoyl-methyl-dithiophosphate. It is yellowish viscous oil or crystalline solid with a garlic like odour, melting point 25-26°C. Its structural formula is as follows :

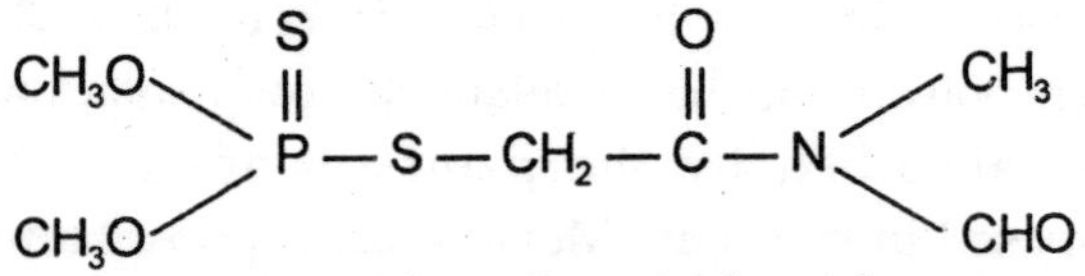

It is a systemic insecticide and acaricide with contact action. Formothion is absorbed by the plant and hydrolysed into such metabolites which are less harmful to mammals but toxic to insect pests. Its acute oral and dermal LD 50 values to rat are 400 and 400-1680 mg/kg respectively. It is formulated as 25% EC and available in the trade name of Anthio.

Thiometon ($C_6H_{15}O_2PS_3$)

Chemically it is O, O-dimethyl S-ethyl mercaptoethyl-dithiophosphate. Thiometon is a dithioate analogue of methyl-O-demeton. Its structural formula is as follows :

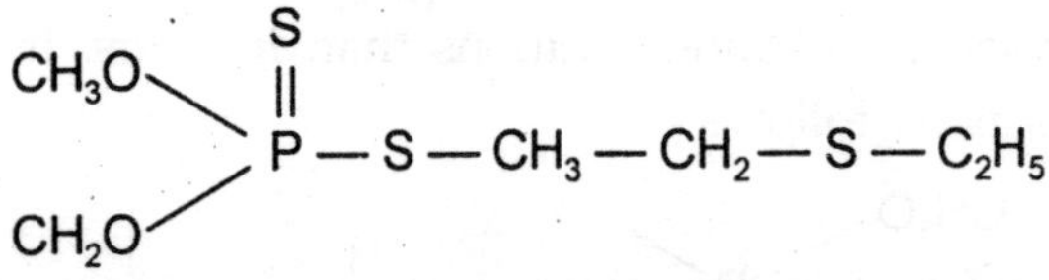

It is a strong systemic insecticide and acaricide effective against sucking insects with particular reference to aphids and sawflies. It is comparatively safer to beneficial and predatory and parasitic insects. It is hydrolysed slowly in plant system by oxidation and saponification and remain effective upto 2-3 weeks. Its acute oral and dermal LD 50 values to rat are 100 and 200 mg/kg respectively. Thiometon is formulated as 25% EC and sold in the trade name of Ekatin.

Phorate ($C_7H_{17}O_2PS_3$)

Chemically it is *o, o*-diethyl-i-(ethylthio) methyl

phosphorodithionate. It is a yellowish liquid, boiling point 118-120°C. Phorate is slightly soluble in water but is miscible with many organic solvents. It gets hydrolysed in the presence of moisture and alkaline conditions. Its structural formula is as follows :

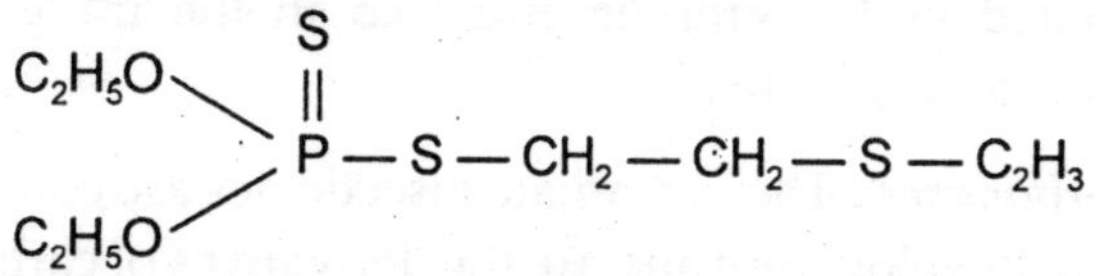

It is systemic insecticide and is absorbed by roots and subsequently translocated to aerial parts of the plant. This material some times causes reduction in stems from seed treatment. It has also shown some nematicidal and acaricidal action. It has been found useful in the control of sorghum shoot fly, aphids, sucking pests of cotton, white grub and gall midge of rice etc. Its acute oral and dermal LD 50 values to rat are 2 to 3 and 70 to 300 mg./kg respectively. It is generally formulated as 10% granule and available in the trade name of Thimet.

Thiodemeton or Disulfoton ($C_8H_{19}O_2PS_3$)

Chemically it is *o*, *o*-diethyl-S-2-(ethylthio) ethyl phosphorodithioate. It is a colourless oily liquid with characteristic odour, Boiling point 62°C. Disulfoton is slightly soluble in water and soluble in most of the organic solvents. Its structural formula is as follows :

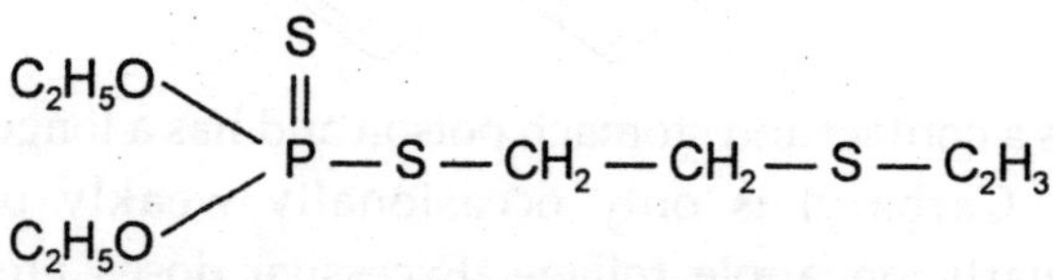

It is oxidised in plant tissues to form the sulfoxide and sulfone metabolities which are responsible for the systemic toxicity of the compound.

Sufficient moisture should be present in the soil for it to act effectively. It is specially effective for preplanting or seed dressing application as a seed treatment to control insects and mites after emergence of the plants. Its acute oral and dermal LD 50 values for rat are 4 and 50 mg/kg respectively. It is formulated as 5% granule and sold in the trade names of Disyston and Solvirex.

Carbonates : The carbamate insecticides are comparatively recent in development and are the derivatives of carbamid acid and have an OCOn = group in the molecule. They chiefly act by inhibiting the cholinestrase enzyme of the neuromuscular system. The commonly used insecticides of this group are as follows :

Carbaryl

It is I-naphthyl-N-methyl-carbamate and available in the name of Sevin. Carbaryl is white crystalline solid, m.p. 142°C. The technical material is 95% pure. It is very slightly soluble (0.004%) in water and moderately soluble in most organic solvents. It has the following structural formula :

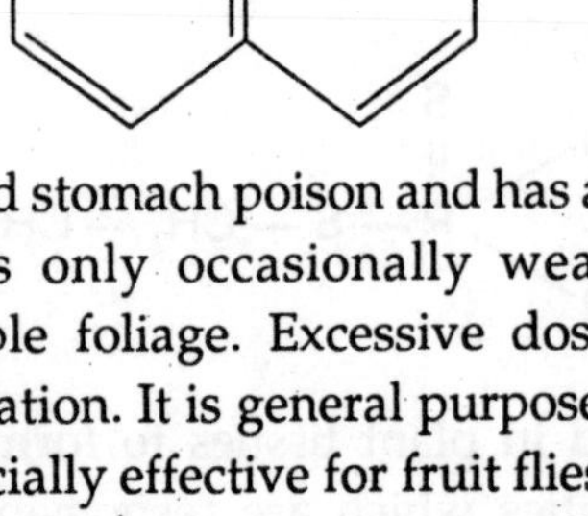

It is a contact and stomach poison and has a longer residual action. Carbaryl is only occasionally weakly phytotoxic particularly on apple foliage. Excessive doses are likely to reduce seed germination. It is general purpose insecticide and has been found specially effective for fruit flies and borer pest. Its acute oral and dermal LD 50 to rat are 400 and 7500 mg/kg respectively. It is formulated as 5% to 10% dusts and 50%

wettable powder. Sevimol 40 LV is a product containing 40% carbaryl with molasses. Granule containing 4% in each of carbaryl and gamma BHC is formulated with the name of Sevidol.

Isolan

It is a product containing l-isopropyl-3-methyl-5-pyrazolyl dimethyl-carbamate. Isolan is a colourless liquid boiling point 105°C to 107°C. It is miscible with water and soluble in most organic solvents. Its structural formula is as follows :

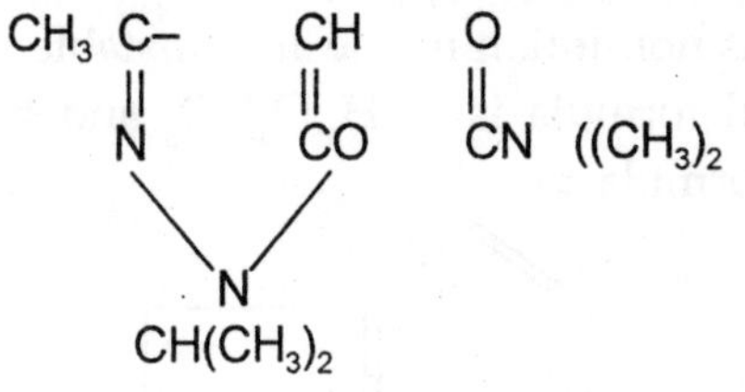

Isolan is a systemic insecticide and also acts as contact, stomach and fumigants. It is mainly used against aphids and in fly baits. Its acute oral and dermal LD 50 values to rat are 12 and 35-60 mg/kg, respectively.

Aminocarb

It is the common name for the product Metacil having 3-methyl-4-dimethylanimophenyl-N-monomethyl carbamate. Aminocarb is a white crystalline odourless substance, m. p. 93-94°C. It is slightly soluble in water, moderately soluble in aromatic solvents and readily soluble inorganic solvents. It is unstable in strongly alkaline medium. Its empirical formula is $C_{11}H_{16}O_2N_2$ and structural formula is as follows :

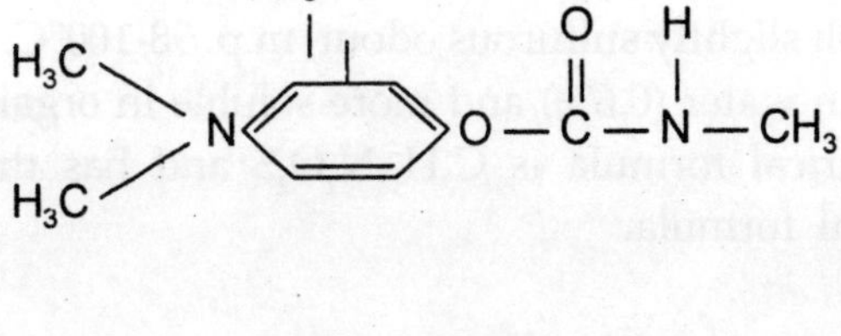

Aminocarb is effective against biting insects with particular reference to Lepidop-terous larvae and also some acaricidal properties. Its acute oral and dermal LD 50 values to rat are 30-50 and 275 to 1000 mg/kg respectively. It is generally formulated as 5% dust and 50% wettable powder.

Carbofuran

Chemically it is 2, 3-dihydro-2-2dimethyl-7-benzofuranyl methylcarbamate. It is a white crystalline solid with slightly phenolic odour, m.p. 153-154°C. Carbofuran is slightly soluble in water (250 to 700 K ppm) and soluble in most organic solvents. It is non-inflammable and unstable inalkaline media. Its empirical formula is $C_{12}H_{15}O_3NO_3$ and has the following structural formula :

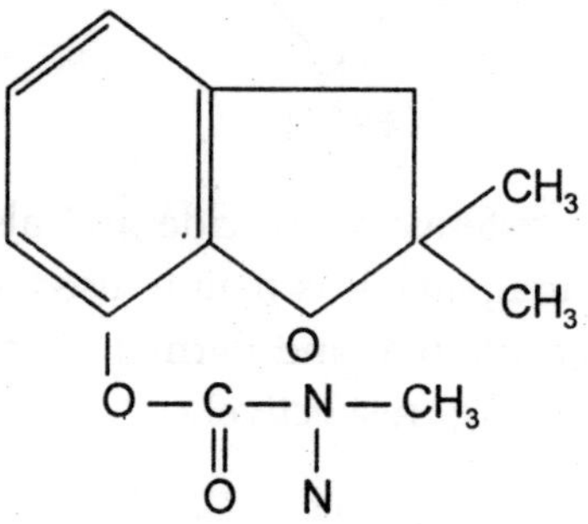

Carbofuran is systemic insecticide and effective against sucking pests, thrips, mites, nematodes and other soil inhabiting pests. Its acute oral and dermal LD 50 values to rat are 8.2-14.1 and 10,200 mg/kg respectively. Its trade name is Furadan and formulated as 3% granule and 50% S.P. or F.P.

Aldicarb

Chemically it is 2-methyl-2-(methyl thio) propionaldehyde 0-l-(methyl carbamoyl) oxime. Aldicarb is white crystalline solid with slightly sulfurous odour, m.p. 98-100°C. It is slightly soluble in water (0.6%) and more soluble in organic solvents. Its empirical formula is $C_7H_{14}N_2O_2S$ and has the following structural formula.

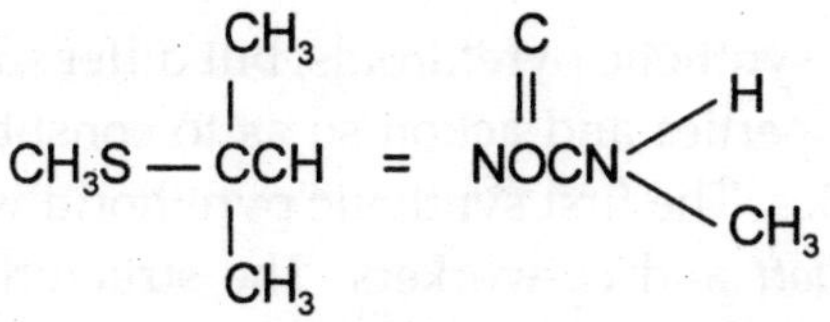

Aldicarb is a systemic insecticide, acaricide and nematicide, effective against a wide variety of insects, mites and nematodes. After application in the soil, it is rapidly absorbed by the plants and oxidised to the principal toxicant aldicarb sulfoxide which is further oxidised to the less toxic aldicarb sulfone. Its acute ora and dermal LD 50 values to rat are 6.2 and 3200 mg/kg respectively. It is formulated as 10% granule and available under the trade name of Temik.

Primicarb

Chemically it is 2-dimethylamino-5, 6-dimethylpyrimidinyl dimethyl carbamate. It is fast acting selective aphicide of moderate mammalian toxicity. Primicarb is not miscible with water but very soluble in most organic solvents, m.p. 90.5°C. It is colourless, crystalline solid, non-corrosive to the material normally used for spraying equipment. Its empirical formula is $C_{11}H_{18}N_4O_2$ and has the following structural formula.

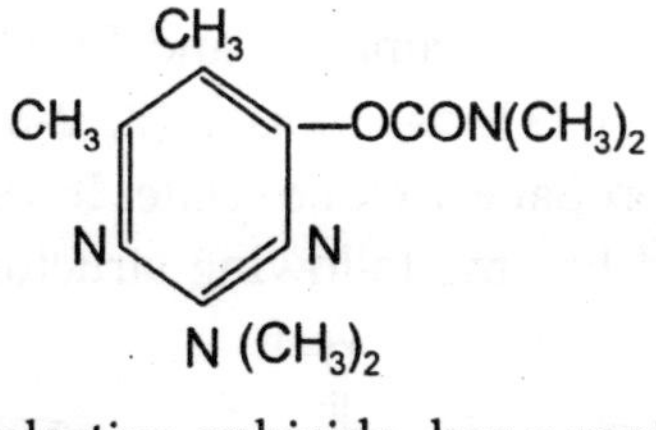

It is a contact selective aphicide, has a rapid knock down and high activity at low concentrations and also act as a fumigant with no unpleasant odour. Its acute oral LD 50 value to rat is 147 mg/kg. It is formulated as 50% dispersible powder and available under the trade name Pirimor.

Synthetic Pyrethroids

The synthetic pyrethroids are mostly based on the earlier

natural and synthetic pyrethroids, but differ so markedly from them in properties and action so as to constitute a new class of insecticides. The first synthetic pyrethorid was described in 1973 by *Elliott* and co-workers. The structure of the natural pyrethrins, of which pyrethrin I is the prototype, has been modified to give successive synthetic compounds with improved properties. The following are the characteristics of these insecticides :

1. They are quick in action.
2. These have high insecticidal efficiency.
3. Comparatively have low mammalian toxicity.
4. Good persistence.
5. Amenability to synergistic action.
6. Small quantity needed to kill the insects.
7. Less risk of environmental contamination.
8. Less chances of insecticidal resistance.

The important insecticides of this group are as follows :

Permethrin

It is also known as ambush and PP 557. Chemically it is 3-phenoxybenzyl (±) *cis, trans* 2-(2, 2-dichlorovinyl)-2, 2-dimethyl cyclopropane-1 carboxylate. Its empirical formula is $C_{21}H_{20}O_3C_{12}$ and has the following structural formula :

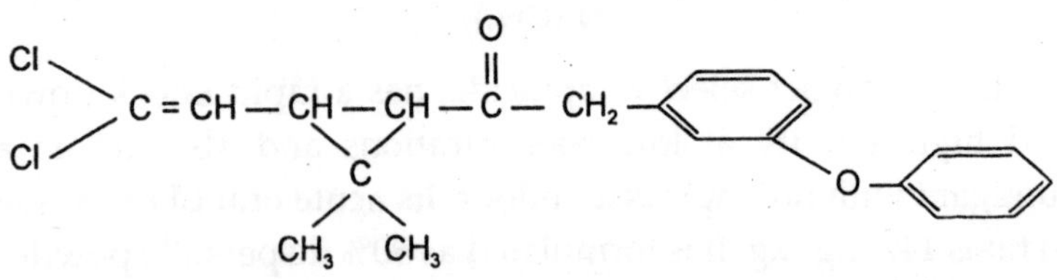

It is practically insoluble (0.2 ppm) in water but miscible with most organic solvents. The compound is both a stomach and a contact insecticide. It is extremely fast acting and has shown adulticidal, ovicidal and particularly larvaecidal activity.

Thus it possesses a very high level of activity against a wide range of insect pests, particularly Lepidoptera. Its acute oral and dermal LD 50 value to rat is 74000 mg/kg. It is formulated as 25 and 50% emulsifiable concentrate.

Cypermethrin

Also known as cymbush and PP 383. Chemically it is cyano-3-phenoxy benzyl (±) *cis. trans*-3-(2, 2-dichlorovinyl)-2, 2-dimethyl cyclopropane-carboxylate. Its empirical formula is $C_{22}H_{19}O_3NCl_2$ and has the following structural formula :

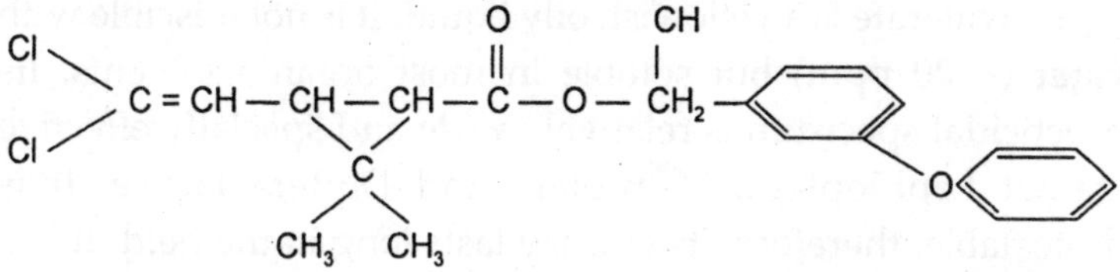

It is practically insoluble in water (0.2 ppm) but miscible with most organic solvents. It possesses a very high level of activity against a wide range of insect pests, particularly Lepidoptera. The activity is usually 10 times greater than that of organophosphates and carbonates. Unlike natural pyrethrins, it is stable to degradation in light, hence, it is relatively persistent on plant surfaces, even under extremes of temperature and sunlight. Its acute oral and dermal LD 50 values to rat are 4030 and 4800 mg/kg respectively. It is formulated as 25% emulsifiable concentrate.

Decamethrin : It is also known as decis. It acts on insects by contact and has a direct insecticidal effect on biologically active instars of pests. It is not broken down rapidly by the enzymes in the insects, and larvae die within a few hours. It is 100, 80, 50 and 40 times more powerful than DDT, carbaryl, malathion and parathion, respectively. Decamethrin shows resistance to leaching out during heavy rain or sprinkler irrigation. The acute oral toxicity to rat is (LD 50) 128.5 to 138.7 mg/kg. It is formulated as 2.8 w/w emulsifiable concentrate containing 25 decamethrin per litre.

Fenvalerate : It is also known as sumicidin and chemically it is cyano-*m*-phenoxy benzyl-isopropyl-*p*-chlorophenyl-acetate. Its empirical formula is $C_{25}H_{22}ClNO_3$ and has the following structural formula :

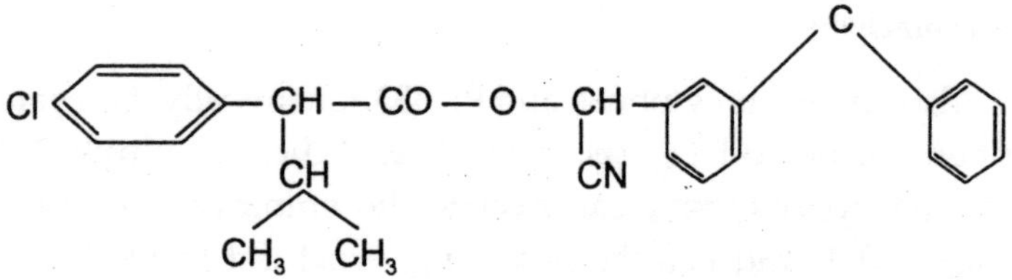

Fenvalerate is a yellowish oily liquid. It is not miscible with water (< 20 ppm) but soluble in most organic solvents. Its insecticidal spectrum is relatively wide and specially effective against Lepidoptera, Hemiptera and Diptera larvae. It is photostable, therefore, its efficacy lasts long in the field. It has quick killing and knockdown effects. The acute oral and dermal LD 50 values to rat are 541 and 73200 mg/kg, respectively. It is formulated as 20% emulsifiable concentrate.

Fumigants

Gaseous poisons use to kill insects are called as fumigants. The application is generally restricted to plants or products in tight enclosures or to soil. They may be used to control all types of insects with irrespective of types of mouth parts since the gas enters the insect body through the spiracles during respiration.

The fumigant which vapourises readily at room temperature is the most useful. For soil fumigation, however, the slower release of vapour from substances, boiling at high temperature has proved most effective. An ideal fumigant is selected by its safety to human beings, animal and plants; penetration quality; non inflammable; harmless to food stuff; non-corrosive to metals; have great power of diffusion; cost; effect on germinating seeds and relative effectiveness. The fumigants may be divided into three broad groups :

1. Gases at ordinary temperature, *e.g.*, Methyl bromide, Ethylene oxide etc.
2. Liquids at ordinary temperature, *e.g.*, Ethylene dichloride, Ethylene dibromide and Carbon tetra chloride.
3. Solids at ordinary temperature, *e.g.*, Sodium cyanide; Aluminium phosphide. The gases are usually compressed in cylinders from which they are released as gases of liquid which vapourise readily. Fumigants that are liquid can be sprayed or sprinkled on the commodities or in the enclosed space soil fumigants can be vapourised by heating or simply by mixing them in the commodity. In the operation of any fumigant the first principle to be observed is to safeguard human lives and only trained personnel with necessary protective appliance should be entrusted with the work. The space to be fumigated must be nearly airtight and it is better to avoid windy or cold weather. There should be provision for ventilation after fumigation.

The dosage for fumigation is expressed in terms of kg/ 1000 cu.m. In fumigation, important things is the amount of gas acting on the insects over a certain period of time. Successful fumigation is dependent upon the attainment of a critical value (ct) which will kill atleast 99% of the insect population. The ct value is the product of the concentration of the gas per unit volume x the duration of the exposure (time). Therefore, it is necessary to know the volume of the enclosed space to be treated and the rate of release of the fumigant vapour. It is apparent that the longer the period of exposure, the lower the concentration of gas required. The sorption of the gas by the material fumigated affects the dosage and ct value of the fumigant, and further, the rate of sorption is dependent on the moisture content of the material. The commonly used fumigants are described here as under.

Hydrogen Cyanide or Hydro Cyanic Acid (HCN)

Of all the chemicals used as fumigants for insects, HCN is extremely used inspite of its extreme toxicity to the human beings. It was originally employed for the control of cottony cushion scale in the U.S.A. It is a colourless, volatile liquid boiling point 26°C. and melting point about 14°C with a bitter almond odour. HCN is a gas at room temperature which has sp. gravity of 0.9483, hence is slightly lighter than air. Since the gas is lighter than air, the diffusion is mainly upward and outward. Under normal atmospheric pressure the gas is not able to penetrate closely packed materials and so partial vacuum fumigation is employed. There are four principal methods of applying this gas in any enclosure.

By the pot method : In this method the gas is generated from earthen jars or barrels placed within the enclosure to be fumigated by mixing water and sulphuric acid in the container. When the mixture is still hot, as a result of themical action, the proper amount of sodium or potassium cyanide is added which produces HCN gas. This method is not in much use.

$NaCN + H_2SO_4$-------------$NaHSO_4 + HCN$

The mixture may be used in the following ratio :

(a) For houses and godowns :

Water : 2 lit.

Sulphuric acid : 1 lit.per 30 cu m. space

Sodium cyanide : 1/2 kg

(b) For plants :

Water : 12 ml.

Sulphuric acid : 6 ml.per 30 cu. m.

Sodium cyanide : 4 g

Machine method : Here the measured quantity of H_2SO_4 is introduced to a solution of Na CN in a special chamber or machine. In this way successive charges of gas are generated and lid from the machine into the enclosure to be fumigated through pipes leading from the machine.

Liquid method : In this method, HCN is stored in liquid form in steel containers and is pumped out in the desired quantity through pipes and nozzles to the enclosure to be treated. 13 g liquid HCN is equivalent to 28 g of NaCN and is used @ 20 mg/30 cu.m. of space. This method is generally used in cases where large amount of gas is needed at one time.

The dry method : In this method $Ca(CN)_2$ or liquid HCN absorbed in inert earths or crude paper discs are sold in hermetically sealed cans. HCN is generated by spreading the material over the floor or blowing it into the atmosphere, when it readily gives up free hydrogen cyanide.

$$Ca(CN)_2 + 2H_2OCa(OH)2 + 2HCN$$

$$CaH_2(CN)_4 + 2H_2OCa(OH) + 4HCN$$

Generally two principal types of calcium cyanide are available in the market, one is commonly used in the form of Cyano gas which contains 40-50% CaCN and gives its gas very slowly. Another form of dust is Cal cyanide' which contains 88% CaCN and gives of its gas very rapidly upon exposed to the moist air.

HCN is one of the most deadly gases, therefore, should never be undertaken by inexperienced persons. Proper gas masks should be used.

Carbon bisulphide (CS_2)

It is also known as carbon disulphide and first employed in France to control pests of stored grains. It is a colourless, bad smelling liquid (due to traces of H_2S), slightly soluble in water (0.22 g/100 g at 22°C) and soluble in organic solvents. The gas has a sp. gr. of 2.63 and is highly inflammable and explosive in mixture above 1% in air. The gas is about two and a half times heavier than air, therefore, has considerable penetrative power. It is used for the control of stored product pests in godowns, and pests of cloths etc. Its emulsion is used for controlling soil insects and nematodes. Best result from fumigation with CS_2 to each 100 cub. ft of space or 7-12 lbs/ 370 qtls of grains. It should not be used directly on seeds, but it should be first poured on gunny sacks, rags etc., the latter being kept on the heap of grain. The gas flows down on account of its being heavier than air and penetrates into the heap. It is toxic to green plants, therefore, should not be used for fumigation of nursery material.

To protect against fire hazards, commercial mixture of 20% CS_2 with 80% carbon tetrachloride are available. One such mixture is sold under the trade name of Grain-ocide which is used @ 1.5 to 2 lb/1000 cu. ft of space or 12-20 lbs/370 qtls of grain.

Carbon tetra chloride (CCl_4)

This is thin, colourless, oily compound with 76°C boiling point. The gas has a pungent odour and is not inflammable or explosive. It is almost insoluble in water but is soluble in organic solvents. It is slow acting fumigant of comparatively low toxicity to insect and hence is not used alone but mixed with other much more toxic gases such as ethylene dichloride, carbon bisulphide to make them less explosive and reduce fire hazard. It increases the volatility and distribution of fumigants like methyl bromide, ethylene dichloride and chloropicrin.

Chloropicrin (CCl_3NO_2)

It is also known as trichloronitromethane and is one of the year gases possessing powerful irritating effect on the eyes. At room temperature it is colourless to yellowish liquid, boiling point 112.1°C. The sp. gravity of the gas is 5.7 and, therefore, has remarkable penetrative power. It is not inflammable, corrodes metals slightly but does not bleach or tarnish. It has been chiefly used in flour mills, grain elevators and farmers bins and against soil insects nematodes etc. It should not be used on growing plants or germinating seeds. Chloropicrin is mixed @. 2% with methyl bromide to act as indicator. It is used @ 1 lit./30 cu. m of space.

Ethylene dibromide ($C_2H_4Br_2$)

It is 1, 2-dibromide ethane and is a colourless liquid with boiling point 131.6°C. The gas has a sp. gravity of 6.5 and slightly soluble in water but soluble in most organic solvents. EDB is odourless non-corrosive and non-inflammable liquid available as 5-20% solution. As a soil fumigant it is used for the control of wire worms, white grubs and nematodes @ 18 to 72 kg/hect. It is toxic to many plants, therefore, the treated soil should thoroughly be aerated before planting. It is highly toxic to warm blooded animals and can be absorbed directly through the skin. It is formulated as liquid, granule and even as capsule.

Ethylene dichloride (CH_2ClCH_2Cl)

It is 1, 2-dichlroethane and is colourless liquid, very slightly soluble in water with boiling point of 83.7°C. The sp. gravity of gas is 3.4 and is inflammable between 6% and 16% in air. It is slow in its action and death of insects take place from one to three days after exposure. Since it is highly inflammable, therefore, used in mixture after the addition of about one-third (3 : 1) of its volume of carbontetrachloride. The mixture is entirely free from fire hazard either as liquid or as gas. In India it is sold under the trade name of killoptera. The mixture is directly poured upon the material to be fumigated and allowed to vapourise. The mixture could be transported or stored in cans and easy to handle in the case of grain stores. The dosage for fumigation of grains etc. is 4 to 5 kg/100 quintets with an exposure period of 48 to 72 hours. The temperature during fumigation should be above 75°C. Since EDB is soluble in fats and oil, it is not recommended for use in commodities with high oil content. The mixture is injurious to the germination of most seeds at the recommended dosage.

Methyl bromide (CH_3Br)

It is one of the most widely used fumigants because of its low boiling point (4.5°C). Methyl Bromide or Bromomethane is a colourless almost odourless gas at ordinary temperature. It is slightly soluble in water (1.34 g per 100 g at 25°C) but readily soluble in organic solvents and the gas has a sp. gravity of 3. 29. It requires a longer period of exposure as it kills the insect slowly. A dosage of 1.5 to 2 lb/1000 cu. ft. with an exposure period of 48 hours is advocated for stored grain pests. The gas is not inflammable and mixture with air can not be ignited by flame under conditions of ordinary fumigations. Since it is odourless, therefore, chloropicrin @ 2% is mixed with it to serve as an indicator. Bromomethane is useful for the fumigation of green houses, fruits and vegetables, stored food products, household goods and nursery stocks etc. The germination of seeds is not affected, if they are dry.

Naphthalene ($C_{10}H_8$)

It is white crystalline compound vapourises slowly to form a vapour which has a sp. gravity of 4.4. The gas has a pungent odour and is inflammable mixture with air. It is used against clothes moths and carpet beetles etc. Naphtalene is also used to a limited extent as a soil fumigant for wireworms and as a fumigant in green houses to control spider, mites and thrips. It is generally available as moth balls and moth flakes etc.

Parachloro benzene ($C_6H_4Cl_2$)

It is a white crystalline material, vapourises slowly to form a inflammable gas. The gas has a sp. gr. of 5.1. Parachloro benzene is very slightly soluble in water and is soluble in organic solvents. It is used as a soil fumigant for controlling apple wooly aphis and larva of the peach tree borer and as a house hold fumigant and repellent for clothes moth and carpet beetle. It is also used to protect museum specimens from the attack of dermested beetle larvae.

Phosphine or hydrogen phosphide (PH_3)

It is a colourless, poisonous and inflammable gas which has been used as a fumigant for many years to fumigate grains, sacks and bins. The commercial product is available as phostoxin consists of 3 g tablets or pellets of ammonium carbamate mix the aluminium phosphide. Each 3 g tablet produces in the presence of moisture 1 g of phosphine by the following reaction:

$$AIP + 2NH_4OCNH_2 + 3H_2O \rightarrow PH_3 + A1(OH)_3 + 2NH_3 - 2CO_2$$

The CO_2 and NH_3 released by ammonium carbamate prevents spontaneous ignition of phosphine. The final product formed after decomposition is harmless Aluminium hydroxide. The gas is highly toxic to all stages of insect. Fumigation of packed bulk commodities can be carried and in an air-tight godown under gas proof sheet or in the container itself, if it is air tight. It does not affect the germination power of seeds. It is generally used @ I tablet per ton of produce with an exposure period of 5 days.

List of Insecticides registered on a regular basis under Section 9 (3) of the Insecticides Act, 1968 in India

Sl. No.	*Name of Insecticides*	*Sl. No.*	*Name of Insecticides*
1.	Acephate	2.	Alachlor
3.	Aldicarb	4.	Aliethrin
5.	Alphanapthyl acetic acid	6.	Aluminium phosphide
7.	Anilophos	8.	Arrazine
9.	Alphacypermethrin	10.	Aureofungin
11.	Barium carbonate	12.	Benzene hexachloride (BHC)
13.	Benthiocarb (Thiobencarb)	14.	Benomyl

Contd...

Sl. No.	*Name of Insecticides*	*Sl. No.*	*Name of Insecticides*
15.	Birertanol	16.	Bromadiolone
17.	Butachlor	18.	Calcium cyanide
19.	Captafol	20.	Captan
21.	Cartap hydrochloride	22.	Carbaryl
23.	Carbendazim	24.	Carbofuran
25.	Carboxin	26.	Chlorbenzilate
27.	Chlorfenvinphos	28.	Chlormequat chloride (CCC)
29.	Chlorothalonil	30.	Chlorphyriphos
31.	Copper oxychloride	32.	Copper sulphate
33.	Coumachlor	34.	Coumatetralyl
35.	Cuprous oxide	36.	Cypermethrin
37.	Cyfluthrin	38.	Dalapon
39.	Decamethrin/Deltamethrin	40.	Dichlorodiphenyltrichlorethane (DDT)
41.	Dichloropropane and dichloro propenes mixture (DD mixture)	42.	Dichlorovos (DDVP)
43.	Dicofol	44.	Diflubenzuron
45.	Dimethoate	46.	Dinocap
47.	Dioxathion	48.	Diuron
49.	Dodine	50.	Diazinon
51.	Dieldrin	52.	2, 4-Dichlorophenoxy acetic acid (2,4-D, sodium amine and ester salts)
53.	Ethylene dibromide	54.	Edifenphos-carbon tetrachloride mixture (EDCT mixture 3 : 1)
55.	Endosulfan	56.	Ethofenprox
57.	Ethepon/Etofenprox	58.	Ethylene dibromide (EDB)
59.	Ethyl mercury chloride	60.	Ethion
61.	Fenitrothion	62.	Fenarimol
63.	Fenobucarb (EPMC)	64.	Ferithion
65.	Fenvalerate	66.	Ferbam
67.	Fluchloralin	68.	Fluvalinate
69.	Formothion	70.	Fosetyl-AI

Contd...

Sl. No.	*Name of Insecticides*	*Sl. No.*	*Name of Insecticides*
71.	Gibberellic acid	72.	Glyphosate
73.	Iprodione	74.	Isoproturon
75.	Kitazin	76.	Lime sulphur
77.	Lindane (Gamma BHC)	78.	Malathion
79.	Maleic hydrazide (MH)	80.	Mancozeb
81.	Menazon	82.	Metaldehyde
83.	Methabenzthiazuron	84.	Methoxy ethyl mercury chloride (MEMC)
85.	Methyl bromide	86.	Methyl bromide + Ethylene dibromide (1 : 1 and 3 : 1)
87.	Methyl chlorophenoxy acetic acid (MCPA)	88.	Metalaxyl
89.	Metoxuron	90.	Methomyl
91.	Metribuzin	92.	Metolachlor
93.	Monocrotophos	94.	Mone sodium methane arsonate (M/SMA)
95.	Myclobutanil	96.	Methyl parathion
97.	Nickel chloride	98.	Nicothine sulphate
99.	Oxadiazon	100.	Oxyacarboxin
101.	Oxydemeton-methyl	102.	Oxyfluorfen
103.	Para dichlorobenzene	104.	Paris green (Copper acetoarsenite)
105.	Paraquat dichloride	106.	Pendimethalin
107.	Permethrin	108.	Phenthoate
109.	Phenyl mercury acetate (PMA)**	110.	Phorate
111.	Phosalone	112.	Phosphamidon
113.	Pirimiphos-methyl	114.	Profenofos
115.	Propanil	116.	Propetamphos
117.	Propoxur	118.	Pyrethrum
119.	Quinalphos	120.	Sevidol (4 : Carbaryl and Gamma HCH)
121.	Simazine	122.	Sirmate
123.	Sodium cyanide	124.	Streptomycin
125.	Sulphur	126.	Temephos

Contd...

Sl. No.	Name of Insecticides	Sl. No.	Name of Insecticides
127.	Trichloro acetic acid (TCA)	128.	Triallate
129.	Trichlorphon	130.	Tricyclazole
131.	Tridemorph	132.	Trifluralin
133.	Thiometon	134.	Thiram
135.	Thiophanate methyl	136.	Triazophos
137.	Triadimefon	138.	Warfarin
139.	Zinc phosphide	140.	Zineb
141.	Ziram	142.	Hexaconazole
143.	Pretilachlor	144.	Propiconazole

List of Banned Pesticides

Sl. No.	Name of Pesticides	Sl. No.	Name of Pesticides
1.	Dibromochloropropane	2.	Endrin (DBCP)
3.	Pentachlornitrobenzene	4.	Pentachlorophenol (PCP) (PCNB)
5.	Texaphene	6.	Ethyl parathion
7.	Chlordane	8.	Heptachlor
9.	Aldlrin (*w. e.f.* 1.1.1994)	10.	Paraquat-di-methyl Sulphate
11.	Nitrofen	12.	Tetradifen
13.	Methomyl	14.	Phosphamidon

List of Pesticides not Approved for Use

SI. No.	Name of Pesticides	SI. No.	Name of Pesticides
1.	Calcium arsonate	2.	EPN
3.	Azinpho-methyl	4.	Mevinphos (Phosdrin)
5.	Lead arsenate	6.	2, 4, 5-T
7.	Carbophenothion	8.	Vamidothion
9.	Mephosfolen	10.	Azinphos ethyl

Contd...

Sl. No.	*Name of Pesticides*	*Sl. No.*	*Name of Pesticides*
11.	Binapacryl	12.	Dicrotophos
13.	Thiodemetera/Dosi	14.	Fention acetate
15.	Fention hydroxide	16.	Chinomethionate
17.	Ammonium sulphamate	18.	Leptaphos (Phosvel)

List of Pesticides Under Review by Government

Sl. No.	*Name of Pesticides*	*Sl. No.*	*Name of Pesticides*
1.	Alachlor	2.	Benomyl
3.	Copper acetoarsenite	4.	Diuron
5.	Ethyl mercury chloride	6.	Fenarimol
7.	Menazon	8.	Oxyflourfen
9.	Sodium methane arsonate	10.	Calcium cyanide
11.	Thiometon	12.	Triazophos
13.	Tridemorph	14.	Monocrotophos
15.	Ziram	16.	Zineb

Newer Approaches in Insect Control

Pheromones : The term pheromone is generally used to those chemicals which are secreted into the external environment by an animal and that elicit a specific reaction in a receiving individual of the same species. These chemicals have also been known as ectohormones. Pheromones are divided into two groups :

(i) Which gives releaser effect; this impose a more or less immediate and reversible change in the behaviour of the recipient ?

(ii) Which gives as prima effect; this triggers a chain of physiological events in the receiving organisms ?

The latter class, which generally operates through gustatory stimulation by the chemicals and control such phenomenon as caste determination and reproductive control in social Hymenoptera (Ants and Bees) and Isoptera (Termites).

Mode of Action

Releasing pheromones are typically odorous and act directly on the central nervous system of the recipient. General types of behaviour that may be released include trial following, alarm, sexual activity, aggregation, dispersal and territoriality. From insect point of view two classes of pheromones that promote aggregation will be important.

Sex pheromones : Which are released by one sex only and trigger behaviour patterns in the other sex that facilitate mating etc. These are also known as sex attractant or sex lures, *e.g.*, gyptol.

General aggregation pheromones : Which may be released by only one sex but cause approach responses by individuals of both sexes of the species.

Uses of Pheromones in Insect Control

In insect control only the sex pheromones are used. Sex elicited by insects have been isolated and identified chemically and they are being artificially synthesised in the laboratories of many advanced countries like U.K., U.S.A., etc. Sex pheromones thus prepared are available in the form of pheromone dispensers of several types. They are species specific and separate dispensers are required for separate insects. These dispensers are fitted into the pheromone traps which are of different types. The sex pheromones of female insects are available in the synthetic forms of pheromonal materials (dispensers) which attract the males of same species for mating. The purpose and use of pheromone traps are summarised below :

Monitoring of insect pests : The most important use of pheromone traps is for monitoring purposes. The outbreak of particular insect species may be ascertained at very early stage of its infestation and thus their population may be controlled at very appropriate time. In case of gram pod borer, Heliothis armigera, 4-6 moths per trap per night, warrants the suitable control measures. In this type of trapping (monitoring) four to six pheromone traps/hectare are recommended.

Mass trapping and destruction : Male moths may be captured in large numbers with the help of pheromone traps and thus their population explosion be brought down to considerable level, economically. Mass trapping should be done on campaign basis. Usually, 6-9 traps are used in one hectare for such purpose.

Mating disruption technique or male confusion technique: In this technique the high concentrations of pheromonal material is used so that percipient males become confused and there becomes mating disruption between males and females. Sometimes the antennal (the organ of pheromone perception) loose the power of pheromone perception due to some high concentration of pheromonal material. Thus the male copulates with some unwanted plant materials. In this technique some micro encapsulated material is also used which is dispersed with cultivated field through the helicopter or aeroplanes. Thus there is inhibition of mating and so the multiplication of insect pests is checked to a some extent.

Insect survey and collection : Pheromone traps are also instrumental in the survey and collection of a particular insect species in the targeted area. It does not require the technical skill and expertise for the identification and collection of desired species.

Lure : Is the mechanism through which insects are directed through food mating/oviposition sites and aggregating. It operates through specific chemical (S) entitles, *e.g.*, methyl euginol, floral scent, carbon dioxide, lactic acid, sugar and propionitrol.

Hormones : In recent years hormones are also successfully used in minimising the pest population. The prime candidate for developing hormonal pesticides is the juvenile hormone that all insects secrete at certain stages in their lives. This hormone is secreted by corpora allata which is in the form of two tiny glands in the head. Ecdysone is another hormone, secreted from thoracic gland which is responsible for pupation and maturation in insects. These hormones have been shown to change the course of development in insects abruptly when applied at appropriate time and in turn these may be used as pesticides. Juvenile hormone increases the larval instrar when applied which may never pupate; besides, the chemosterilization property of this harmone has also been reported. By the application of Ecdyson hormone insect could be killed or may be brought in developing conditions during wrong season thus committing ecological suicide.

Antifeedant : The idea that a chemical could prevent feeding is not new, early workers usually referred to such materials as repellents. Antifeedants are not repellent in the true sense of the word, since insects are not driven away or kept away. Such chemicals inhibit feeding of the insects and thus protect the crop or commodity and eventuality starve to death without eating the leaves. There are following types of chemicals which show antifeeding property.

Triazene — Compound 24055

Organotins — Breston, Brestenol

Carbamate — Baygon

Botaniical extracts — Pyrethrins, Neem seed etc.

Mode of Action

Insect feeding consists of several stages these are :

(i) Orientation of attraction

(ii) Bitting.

(iii) Swallowing or Sustained feeding.

When insects are given the choice between antifeedant treated or untreated plants, there is little or no difference in initial attractiveness and equal numbers will be found on both groups of plants. Bitting takes place of both groups, but the differents lies in whether, there is a continuous bitting or a sustained feeding response. Those on treated plants continue taking tiny bites from the leaf as they wander around, where as those on untreated plants will feed in one spot unless disturbed.

In order to have sustained feeding it requires :

(i) The presence of gustatory stimulus.

(ii) The absence of any inhibitory stimulus.

(iii) and it must be hungry.

Evidence indicates the 1st of these are the point at which an antifeedant acts. Apparently antifeedant inhibit the taste receptors of the mouth region so that, lacking the proper gustatory stimulus, the insect fails to recognise the treated leaf as food and thus continuous foraging.

Chemosterilants : Chemicals which produce sterility and not killing the pest, are referred as chemosterilants. The basic concept behind the use of chemosterilants is to minimise the reproductive potential of the insects. There are two principles involved in insect control.

(i) Mass production, sterilization and release of sterile organism so that they will interfere with the remaining natural fertile population.

(ii) Sterilization of the required portion of the natural population by appropriate means (baiting etc.) so that the sterilize insects can not reproduce and at the same time they compete with the remaining fertile individuals and check the reproduction.

The sterilizing action of a chemical may be due to the fact that :

(i) It damages the chromosomes and leads to the formation of dominant lethal mutations in the semen of males or mature eggs of females;

(ii) It causes perishing of the cells in the premeiotic stages or aspermia;

(iii) It inactivates the sperm;

(iv) Under the influence of the toxicant, females lose the ability of laying eggs.

The practical application of chemosterilants depends on the following three points ;

(i) Nature of the active compounds.

(ii) Mode of administration.

(iii) Diversity of insect.

Types of Chemosterilants

Chemosterilants may be divided into three groups :

Alkylating agents : Specially effective as the male sterilants.

(a) Aziridines and Non-aziridines, *e.g.*, Tepa, Metepa, thiotepa, apholate.

(b) Sulfuric acid esters.

(c) Nitrogen mustards.

Antimetabolites : Mostly female sterilants, e.g.,

(a) Analogues of purines, pyrimictines.

(b) Folic acid.

(c) Analogues of amino acids.

Miscellaneous **:** Compounds

Examples **:** Hempa, colchicine, caomarin.

Mode of administration **:** Chemosterilants may be applied in the following manner :

(i) Direct contact with sprays and dusts.

(ii) Combination of insecticide and chemosterilant.

(iii) Oral ingestion.

(iv) Sterile male release method.

Advantages

The application of chemosterilants has following advantages:

Bonus effect **:** If the sexual competitiveness of an insect is maintained the fertile organisms will not be incapable of reproducing but have the ability to reproduce the reproductive capacity of the normal unsterilized population.

Space effect **:** The sterile individuals through movement can effect the reproductive potential of the individuals outside the treated area.

Time effect **:** The insects which are long lived and sterilized, they can effect the reproductive potential of the population in subsequent generation by mating competition.

Population crush **:** The sterilized individuals will compete with unsterilized individuals for food and shelter in the environment during their normal surviving period. They also mate with even the hiding individuals.

2

Effective Measures

The use of pesticides has increased considerably throughout the world over during the last three decades. With the need for increased production without increasing the area of crops, more farmers are using improved varieties of seeds, fertilisers and plant protection chemicals. Pest and disease control is often essential. When we use pesticide, proper handling and care is essential otherwise it leads hazards to human beings and environment.

Pesticides may get entry into body by three routes namely by mouth (oral), through the lungs (inhalation) or through the skin (dermal). Poisoning may result from a single dose of pesticide which referred to as "acute" poisoning or by repeated in take of small quantities of poison *i.e.,* chronic poisoning.

Based on acute toxicity data, usually for rats and rabbits, pesticides are often grouped into arbitrary categories of hazard and for case of reference, an example of this type of classification is given in table.

Level of Hazard	*Pesticides*	*LD 50 in rats mg/kg of body weight*		*Colour of Triangle Container*
		Oral	*Dermal*	
Extremely Hazardous	Temik, Endrin Thimet Dimecron, Dieldriin, TEP	Less than 50	Less than 200	Bright red
Very Hazardous	Aldrin, Fenthion Toxaphene, DDVP DDT, Thiodan Rogor yellow Sumithion, Chlordane, Lindane Diazinon	51-500	201-2000	Bright
Moderately Hazardous	Sevin, Malathion, Pyrethrin, Orthene	501-5000	2001-20,000	Bright Blue
Relatively Safe	Sulphur, Methoxychlor	More than 15,000	More than 120,000	Bright Green

Safety precautions should be taken when applying pesticides of different categories of hazard. Improper handling of pesticides may result in fatality or lethal effect to a person. Protective clothing, rubber gloves, shoes and large handkerchief are very much essential to protect the body, hands, legs and face.

First aid should be given when a person using pesticides becomes ill. General symptoms of poisoning are headache, nausea, vomiting, anxiety and in severe case convulsions occur. Immediate treatment such as washing the contaminated skin, flushing the contaminated eye, artificial respiration and vomiting induced with saline solution must be employed for immediate recovery from poison. Immediate medical attention by a Doctor is essential. The doctor must be informed the name of the pesticide and given as much information as possible by showing him a leaflet or label about the chemical so that he can administer effective antidote for immediate recovery.

Other safety precautions for pesticides application are given below :

Before Applying Pesticide-General Instructions

1. Know the pest, and it's extent of damage.
2. Use pesticide only when really needed.
3. Seek advice on the proper method of control.
4. Use only the recommended pesticide for the problem. If several pesticides are recommended, choose the least toxic to mammals and if possible the least persistent.
5. Read the label inducing the small print.
6. Make sure that appropriate protective clothing is available and is used.
7. Commercial operators using large quantities of organophosphate pesticides should visit their doctor and get their blood tested.
8. Check application equipment for leaks, and ensure that it is in proper working order.
9. Check that pesticides are properly placed in locked store. Avoid inhaling pesticide mists or dusts, specially in confined spaces such as the pesticide store.
10. Warn neighbours about your spray programme, specially if they have apiaries (Honey bees).
11. Take only required pesticide for the application from the store to the site of application. Do not transfer pesticides into other containers.

While Mixing and Applying Pesticides

1. Wear appropriate protective clothing. If it is contaminated remove and replace with clean clothing.
2. Never work alone when handling the most toxic pesticides.

3. Never allow children or other unauthorised persons for this job.
4. Re-check the instructions on the label.
5. Avoid contamination of the skin, specially the eyes and mouth. Formulations should be poured carefully to avoid splashing. Avoid powder formulations puffing up into the face. If contaminated with the concentrate, wash immediately.
6. Never eat, drink or smoke while mixing or applying pesticides.
7. Always have plenty of water available for washing.
8. Always stand upwind while mixing pesticide.
9. Make sure that pesticides are mixed in the correct quantities.
10. Avoid inhalation of chemical, dust or fumes.
11. Start spraying near the downwind edge of the field and proceed upwind so that operators move into unsprayed areas.
12. Never blow out clogged nozzles or hoses with your mouth.
13. Avoid spray drift and do not spray if wind conditions cause drift, otherwise birds, bees and wildlife may be endangered. Never spray if the wind is blowing towards grazing livestock or regularly used pastures.
14. Never leave pesticides unattended in field.
15. Provide proper supervision of those assisting with pesticides application and have adequate rest periods.
16. Do not work with pesticides if your cholinestrase level is below normal.

Instructions after Application

1. Return unused pesticide to the store.

2. Safely dispose of all empty containers. Since it may be difficult to bury empty containers after each day's spraying operations, hence they should be kept in the pesticide store until a convenient number is ready for disposal. It is absolutely impossible to clean out a container sufficiently well to make it safe for use, for storage of food, water or as a cooking utensil. If any container is burnt, never stand in the smoke.
3. Never leave pesticide in application equipment. Clean *equipment and return it* to store.
4. Remove and clean the protective clothing.
5. Keep a record of the use of pesticides.
6. Do not allow persons *to enter the treated areas for the required period if* restrictions apply to the pesticide used.

Treatments in Case of Poisoning

In case of pesticidal poisoning to a person *following treatments should be followed* immediately :

1. The efforts should be made to remove the poison from the body of affected person.
2. Antidote should be administered as early as possible.
3. The excreta etc. should be collected and sent for further examination.
4. Arrangement should be made to carry patient to some hospital.

In Cases of Swallowed Poison

In case of swallowing the poison, efforts should be made for vomiting and stomach may be washed with stomach tube. For inducing vomiting, following substances may be given :

1. Saline solution (15 g Common salt in a glass of warm water).

2. One gram zinc sulphate in a glass of water.
3. Some quantity of soap in a glass of warm water.
4. 1/10 gram injection of apomorphine hydrochloride.

If the patient is unconscious, then vomiting should be avoided and help of the doctor must be taken immediately and after washing the stomach some antidote may be given. It is always better to give universal antidote which contains following ingredients:

(a) Charcoal — 2 parts
(b) Tamic acid — 1 part
(c) Milk of magnesia — 1 part

Some other Common Antidotes are as Follows

1. Atropine sulphate
2. BAM (Diacetyl monohexene)
3. PAM (Pyridine aldoxymet)

In Case of Inhalation of Poison

Following case should be observed in case of inhalating the poison :

1. The patient must be brought immediately in open air.
2. If he is in the house, all the ventilators and doors should be opened.
3. The cloths of the body may be loosen.
4. The patient must be saved from cold and may be wrapped with blanket.
5. If he is feeling trouble in respiration artificial oxygen may be employed.
6. In case of headache, convulsion and unconsciousness patient should be kept in dark places and avoid noise.

Antidotes for the Treatment of Pesticide Poisoning

Pesticide	*Symptoms of Poisoning*	*Antidotes*
Organochlorine Aldrin, chlordane, heptachlor, DDT, HCH, lindane, dicofor, endosulfan	Nervousness, headache, dizziness, nausea, vomiting, tremors and diarrhoea	(i) Phenobarbital 0.7 g per day or pentabarbital 0.5 g per day (ii) Do not give morphine, theophylline or aminophylline
Carbonates Carbaryl, aldicarb, thiobencarb, carbofuran, propoxur .	Giddiness, headache, nausea, diarrhoea blurred vision muscular twitching, convulsions	(i) Inject atropine sulphate 2-4 mg. for an adult (0.04-0.08 mg/kg body weight for children for 24-48 hours. (ii) Convulsions and anxiety can be treated with 5.10 mg of diazepam injected intratnuscularly. (iii) While keeping the patient fully atropinised administer also an oxime, if available, *e.g.*, 2-PAM 1000-2000 mg IM or IV for adults or 15 mg/kg body weight for children. Repeat after 1-2 hours.
Organophosphates Malathion, methylparathion, acephate, monocrotophos	Same as carbonates	Same as carbonates
Nicotine	Neurological manifestations	Phenobarbitene sulphate 0.7 g daily
Arsenic compounds	Severe gastrointestinal	BAL (Dimercal prol) 3-5 mg/kg body weight

In Case of Skin Contamination

It is just possible that during the handling of pesticides some parts of the body may come in the contact of poison. Under such circumstance following treatments may be followed:

1. The contaminated part may be washed with soap immediately.
2. The related part of the body should be washed for sometime with running water.
3. If necessary, Burnol may be used to avoid irritation.

Insecticidal Machinery

For the effective control of pests infesting crops in fields or grains during storage, different types of pesticidal machineries are used and their selection mainly depends upon habit and habitat of the pest, type of the crop and its area to be treated etc.

The efficiency of machinery is reflected by the ability to cover maximum area and mortality of insects with minimum use of the chemical. This is possible only when uniform spray is discharged by the equipment and it covers larger area. In order to meet out these requirements, different types of machineries have been designed and developed. They may be grouped as follows :

(i) *Sprayers :* They are mainly used to spray the fluid.

(ii) *Dusters :* They are meant for dusting the chemicals.

(iii) *Fumigator :* By and large they are helpful in fumigating the soil or food commodities.

The various pesticidal application methods and equipments in use have conveniently been classified and depicted in the following chart.

Classification of Machinery

- SPRAYERS
  - Hand operated
    1. Automizer
    2. Slide pump
    3. Compressed air sprayer
    4. Knap sack
    5. Foot sprayer
    6. Rocker sprayer
    7. Stirup sprayer
    8. Charge pump
  - Power operated
    1. Hydraulic power sprayer
    2. Blower sprayer
- DUSTERS
  - Hand operated
    1. Plunger duster
    2. Blow duster
    3. Rotatory duster
  - Power operated
    1. Power duster
- FUMIGATORS
  1. Cyno gas pump
  2. Soil injector

Sprayers

The liquid formulation of insectides are used with the help of a sprayer and the qualities of a good sprayer are given below:

1. It should discharge uniform spray on the treated material.
2. The metal used for the construction of a sprayer should not react chemically with the insecticide and must be strong enough.
3. The construction should preferably be simpler so as to clean it without any difficulty.
4. Sprayer should treat maximum area in minimum time.
5. Solution must be broken down into very fine uniform droplets before it reaches the materials to be treated.
6. Spray must cover the desired distance and height.
7. There should not be any wastage of chemical either through leakage or falling from the sprayer when using it in field.

The description of some commonly used hand operated sprayers is given below :

Hand Automizer (Flit Pump)

It is generally used to kill mosquitoes, houseflies etc. by spraying the chemical in houses. This type of automizer does not give a continuous spray, but spray fluid comes out only when it is pumped like a syringe. It has following parts :

(a) *Barrel :* It is hollow cylindrical and tin made barrel which measures 25 cm in length with a diameter of 4 cm. It is connected of the cap of tank on the lower side of the anterior end and this end of the tube is closed with a small hole in the middle, which is called nozzle. There is a tube (1/8" diameter) projecting from the tank and ending near just half of the nozzle and it travels upto the bottom of the tank and force the liquid in the spray form. The barrel has a small air hole (1/6" diameter) on its posterior end and it acts as an outlet and inlet for the air.

(b) *Barrel cap :* The posterior end of the barrel is also covered with a lid known as barrel cap which has a hole in its centre through which plunger rod moves.

(c) *Plunger rod :* It is made up of a wire (1/6" diameter) and provided with a wooden handle on the posterior end which helps in the operation of the plunger. On the anterior end of plunger, a washer and a nut are attached and in between, there is one leather valve, which closely fits within the walls of the barrel.

(d) *Tank :* It is rounded in shape and have the capacity of half pound. The tank is furnished with a cap known as filler cap.

Working : It operates almost in the same fashion as the cycle pump. First of all the liquid to be sprayed is filled in the tank. In the upstroke, when the plunger is pulled back, the air comes in the barrel through the nozzle and when plunger is pressed forward the pressure is exerted on the air. Since the leather valve is directed forward and is closely fitted within the barrel, the half portion of the nozzle is surrounded by the delivery tube. Therefore, the air of the delivery tube also goes out with the air of barrel and thus resulting in vacuum in the delivery tube. This is occupied by the spray liquid which in turn is thrown out in the form of a fine mist.

Hand Compression Sprayer

This type of sprayer is generally used to spray the field crops and trees. It is operated by a single man and it functions like a stove or petromax. This machine is available in two, three and four gallons capacity. By and lage the sprayer has the following important components :

(a) *Pump :* This pump operates exactly on the same principles as a football pump. It is fitted in between the upper end of the tank and hangs inside it. The pump has handle, plunger and plunger spring which are visible outside of the tank, where as barrel, leather valve and ball valve are found inside the tank. The function of the ball valve is to check the incoming of the air which is filled in the tank back into the pump.

(b) *Tank :* It has the cylindrical shape with the upper end convex and is made of brass or copper. The pump is fitted in the middle of the tank and the spray liquid is filled through a small opening having a diameter of 1 inch and covered by a filter cap which is situated on the upper end of the tank. The pressure gauge is attached just close to it which is meant to determine the pressure of air inside the tank. The bottom of the tank end is connected with other tube. The former tube is called delivery tube, this tube runs parallel to the bottom where as the latter one is a lance. In order to check the flow of the spray liquid by the operator, there exists one cut off cock between the joint of tubes. Besides, there is one stand that keep the tank in a right position and it is provided with a waist-plate. On the upper side of the tank, there is one circular plate around its body which is fitted with buckets. Each bucket is attached with the strap and at the distal end of each strap the hooks are provided. During spraying operation these hooks are connected with the waist plate.

Working : Before spraying operation, the liquid is filled in the tank preferable upto 3/4th of the total capacity of the tank. Thereafter the pressure is developed by pumping to the level of 55-60 psi (per square inch). Following this, sprayer is fitted on the back of the operator with the help of buckets, straps, hooks and wast plate. Then the cut-off cock is opened which is just near the bottom of the tank to fill the spray liquid in the form the spray of the liquid, the other cut-off cock which is fitted at the base of the extension rod is opened and thus the spray liquid is thrown out through nozzle in the form of a fine mist by the force of air pressure. In this type of machine the spraying is continued as long it is desired.

Paddle Pump Sprayer

This type of sprayer is more suitable to the field crops and particularly the trees which are upto the height of 20 feet.

An iron stand is provided as base to this sprayer. This machine is the assembly of a plunger, a stand suction hose, delivery hose and as extension rod having a spray nozzle at its free end. Since this is a foot operated sprayer, a paddle is attached to the plunger rod to make its up and down movement which ultimately operates the sprayer.

Working : No tank is provided to this sprayer but a separate bucket containing spray liquid works as a tank. A suction hose which is generally 2 m long and having a strainer at its free end is dipped in the spray liquid when the pedal is pushed down with the foot. The plunger is pushed up in the body of pump thus the air is compressed and moves out of the delivery tubes. The vacuum thus created results in the filling of the suction tube with spray liquid. Continuous paddling is required which sucks the liquid mixture into the body of the sprayer and is released through the delivery tube to the lance and finally through the nozzle which can be adjusted for desirable mist. One sprayer with a nozzle is enough to spray one hectare of a medium sized crop in day.

Power Sprayers

The manually operated sprayers are used for small holdings (up to 4 hectares) and for large scale spraying, power sprayers are required. Broadly, they may be classified in to three groups:

(i) Engine mounted power sprayer.

(ii) Tractor mounted power sprayer.

(iii) Aeroplane sprayer.

On the basis of amount of liquid sprayed and concentration of insecticide, they may be classified as follows :

(i) Low concentration and high volume sprayer.

(ii) High concentration and low volume sprayer.

(iii) Mist blowers and fog or aerosols generator.

Low concentration sprayer : This is the simplest form of power sprayer in which power is generated by a petrol or diesel engine of one or more cylinders and may have the capacity of 10 to 270 litres per minute. There is provision for overflow with a mechanical arrangements for agitation of spray liquid in the tank. The air chamber has a sturdy structure and helps in maintaining a high pressure of 25 kg per sq. cm. The engines are generally of 1 to 20 H.P. depending on the pump's capacity and pressure range. They draw the power is taken off the tractor. In such type of sprayers the spray liquid is forced out through the nozzle in small droplets under pressure developed by the pump. The droplets are deposited on the surface sprayed by the sheer force of the spray. The delivery hose varies from 25 m to 103 m or more in length and is provided with a hand operated spray gun which has a control valve mechanism for adjusting the spray pattern. In these sprayers the concentration of the spray liquid is low, quantity of spray used per hectare is about, 1,120 litres and the particle size is big.

High concentration low volume : In these sprayers a relatively small volume of liquid is automised by a large volume of air high velocity. The principle involved is the same as in the case of hand automizer. They produce a much greater degree of automization then the nozzles used in low concentration sprayers as the spray liquid is discharged at 7 kg per sq. cm. These sprayers work by compressed air or have a rotary fan and a propeller turbine for production a current of air blowing at very high velocity into which a small quantity of spray material is introduced by gravity or with the help of a high pressure pump. In this case only 10-110 litres of the spray liquid is necessary for one hectare. Special nozzles are required for these sprayers and generally the insecticides are used in E.C. form because wettable powders may clog the nozzles.

Mist blowers : These sprayers work on the same principle as the low volume high concentration sprayers ; the only difference being that the droplets are broken into particles of small size ranging from 50 to 100 micron.

The fog or smoke generators have been developed to disperse the spray material in extremely fine droplets (1 to 50 microns in diameter). The setting of aerosol droplets is very slow and as such is very much affected by the prevailing wind currents. This type of equipment can be successfully used where the foliage is dense and where fog takes time to settle and does not drift with the wind. A fog machine consists of tank with or without the engine and thermal or mechanical arrangements to break up the liquid into the desired droplet size and to create a force for carrying the fog from the machine.

Dusting Machinery

Hand Duster

This works on the simple principle of put forth the dust of air in which the dust particles are suspending. This type of duster is provided with a glass bottle and a piston to push the air into the duster. The dust is filled in the glass tube which comes out from the delivery hole with the compressed air. The use of this duster is limited to the small scale research plots or kitchen gardens.

Hand Rotary Duster

The appearance of this duster is larger as is operated by fitting it to the left side of the operator.

It is basically made up of a blower, gear box and a hopper. These are of hand carried type of shoulder mounted type of dusters.

Hopper : This is a drum shaped container which has the capacity generally of 5 kg. The hopper should not be filled more than three fourth of its capacity. A brast is attached on the back side of the hopper. Two banded iron bars known as agitator are connected with a crank shaft which passes through the centre of the hopper. These agitators move the dust continuously inside the hopper on the fan side which propels the dust into the blower through the exit hole situated in the hopper. To regulate the discharge of dust a feed regulator is provided on the outer side of this exit hole.

Gear box : There is a box like structure attached to the blower on the right hand side of the drum and consist of four gears. Crank shaft is the main attachment to which the largest gear A is fitted in the centre of the box. This gear rotates a small double gear B with its grooved connections. Similarly, gear B rotates another double gear C and finally to smallest gear D which is attached to the fan shaft. This gear system helps in causing many rounds of gear D for one round of gear A and thus fan moves very fast. To operate the main gear one banded iron rod is linked with the crank shaft and called as crank.

Blower : This is a closed round shaped structure which opens to the opposite side of the drum through a small projected opening generally 1". in diameter. A fan provided with five blades arranged longitudinally is fitted in the centre of the blower. Rotation of this fan throws the air outside to create vacuum in the blower which is further occupied by the dust from the hopper and air from wire netting provided on the left side of the blower. The projected end of the blower is connected with a long tube known as lance which is fitted with a triangular reflector at its distal end.

Working : Dust should not be filled more than the 3/4 of the capacity of the hopper and this mist be closed tightly with its lid. The duster is tied to the left side shoulder of the operator with the help of straps provided with the duster keeping the breast plates on the chest. Lance is attached to the blower properly and reflector is also adjusted according to the need while operating the duster. Operator should move slowly forward in the direction of the wind rotation the handle continuously to ensure the uniform dust coverage of the crop.

There are other types of dusters also namely plunger type, blower type, knapsack type, fan type and power dusters, but they are not very common in use.

Power Dusters

These are operated by means of an engine or a tractor power take off device. A power duster comprises a hopper containing one or more agitators, a regulator for controlling the discharge of dust, a fan or blower producing an air blast for drawing in the dust from the hopper and blowing it out and flexible pipes with nozzles. The dust regulation device adjusts the dust flow at desired rate ranging from 1 to 10 kg per minute. The machines have arrangements for mixing a blast of air current with the powder. The mixed stream blowing out at velocity of about 320 km per hour.

The fans provided are mostly of centrifugal type. The fan case provided with one outlet or a series of four to eight or more outlets. These outlets are connected with flexible pipes to nozzle arranged on a boom in the rear of the tractor for carrying the dust. A power duster with one outlet can dust 12 to 16 hectares per day, depending upon the crop height and dosage.

Spraying types : Based on the volume of spray mixture applied per unit area (at the rate of lit/hect) through different types of sprayer classified as high volume 750-1000 to be applied through manually operated sprayer, low volume 650-150, to be applied through motorised sprayer and less then 12 lits, to be applied through air craft or specially designed motorised/ battery operated equipments.

Dusting versus Spraying

1. Dust can be used where water is limited or scarce.
2. Dusting is a quicker process as compared to spraying.
3. Sprays are more effective then dusts having the same amount of toxic chemical.
4. Sprays are not affected so much by wind as the dusts.
5. Excess of spray applied simply runs off, keeping safety value against excessive deposits.
6. The cost of dusts is comparatively low than the sprays.
7. The equipments required for spraying are comparatively costlier and heavier than the dusters.
8. In the case of dusting, uniform deposition is not possible, while in case of spraying it may be obtained upto a great extent.

9. Dusting is preferred in morning hours, while the spraying during evening hours.

Fumigators

These are generally used to kill insects or other pests by generating the gas through liquid or solid substances. There are different kinds of fumigators, but the commonly used one is ordinary fumigator cyno-gas pump which is described below:

Cyno-gas Pump

This fumigator is used for buffing the calcium cyanide powder or the cyno-gas. It has the following important parts:

Pump cylinder : It is made up of brass tube which is 18" long and 3" in diameter. The upper end of this tube is closed by a pump cap and its lower end opens into a glass bottle through a small nozzle. This glass bottle is filled with the powder. In between the pump cylinder and glass bottle there is a small valve which allows the air to go into the bottle but checks its entry into the pump cylinder.

On the base of this pump cylinder there is one delivery tube which has got its connection with bottle. The air mixed with powder comes through this tube. Besides, one foot rest is also provided on the lower side of the pump cylinder.

A delivery hose made of plastic is connected with the delivery tube. In the posterior end of the delivery hose one thin brass rod is connected which measures one foot in length. The rod is known as lance.

Plunger rod : The upper end of the plunger is connected with a wooden handle, while the other end has the leather valve which is fitted with washers etc.

Working : Cyno-gas pump is generally utilized in fields and godowns to kill rodents, ants and termites. First of all the bottle is filled up with the powder upto 3/4th of its capacity and there after the lance is inserted after slightly digging the ground of locating the burrows where the rats are present. Following this, the pumping is done and as a result the powder goes inside the burrows. Generally, two ounces of the poison is pumped in each burrow. When the operation is completed the holes should be closed with wet soil or by any other suitable means so that the rats may not come out.

Care and maintenance of plant protection equipments : The following measures may be observed to keep the machines in good working order :

1. Clean the machines properly and store them in dry place when not in use.
2. Flush the equipment with clean water to wash inside parts of containers, tubes and nozzles to keep them free from chemical.
3. The hoses and nozzles of the sprayers should be thoroughly washed with clean water after use and then kept dry. Similarly, the dust from the duster should be taken out and cleaned with a cloth.
4. The moving parts should be greased and oiled regularly.
5. Keep the machines in working order by following a regular maintenance programme.
6. The oil should be kept at the required level in the power equipment.
7. Rubber hose should not be bent at angle when in use and should be removed and kept rolled when not in use.
8. Always stock spare parts, washers, nozzles etc. for replacements when the old parts get worn out.
9. Inspect pump valve seat, plunger cups and packing atleast once in a year to check leakage from these parts.
10. The spray liquid should be strained so as to avoid clogging of the nozzle and the delivery tube.

11. Test pressure regulator and pressure gauge periodically for efficient functioning of the sprayers.
12. Change oil regularly in the crank case to prevent corrosion of the bearings, gears and shafts.

Aerial Spraying

Aircrafts as a means of dispersing agricultural chemicals are now becoming popular as ex-military aircrafts, easily to adopt for agricultural work and ex-airforce pilots, desirous to continue flying, are available. This is supported by the availability of such pesticide which can be sprayed without dilution, making the spraying application economic. An aeroplane was first used in pest control in 1920 and now more than 40 countries are using aircrafts for this purpose. Aircrafts of the power ranging from 90-500 H.P. are used.

Spraying pattern varies with the nature of the chemical and compound-used, therefore the personal dealing with the spraying must be familiar with the term "high volume", low volume and "U.L.V. "etc. United states Department of Agriculture (USDA) has defined these terms as follows :

Full Coverage Spray or High Volume Spray

When used in labelling for agricultural chemicals, the term full coverage spray will apply when the total volume of spray to be applied will thoroughly cover the crop being treated to the point of run off or drip. High volume spray is applicable to low solubility or dispersibility of chemicals or highly soluble product. There is no danger of phytotoxicity but the availability of more labour and supply of water become a problem. Generally the high volume spray techniques require more than 400 litres of spray liquid/hectare.

Low volume sprays : When used in labelling for concentrated or technical agricultural chemicals the term low volume or concentrate spray will apply when the total volume of spray to be applied is adequate to uniformly cover the being treated, but not as a full coverage treatment to the point of run off. Low volume spray techniques are employed where spray liquid volume ranges in between 5 and 400 lit/hect.

Ultra low volume sprayes : When used in labelling for concentrated or technical chemicals, the *term ultra* low volume will apply when the total volume of spray to be applied per acre is 1/2 gallon or less and is to be applied "undiluted". In all the ULV techniques less than 5 lit/hect. spray volume is applied. The various factors decide the choice of best suited techniques such as types of vegetation to be treated, kind of pests involved, availability of water, safety and experience of the operator.

ULV applications have been made generally by the aircraft equipped with a simple modification of the conventional boom and nozzle. For aerial spraying of the insecticides, both fexed-using aircrafts as well as helicopters are used. Correct nozzle tips and rotary automizers, their proper placement on the boom and maintenance of desired flying altitudes are the essential points in ULV application. The volume of insecticide used in ULV technique is extremely low and even a slight error in the above factors may result in inadequate or faulty coverage. At the normal operating speed of aeroplane flying at the height of five to seven meters a pattern of droplets 60 to ISO in diameter is laid down at densities of 15 to 20 meters, while crosswing of 8-11 kph are permissible. ULV application has proved its benefit for the treatment of large agricultural crops. ULV application is more faster thus considerably economic also. Another significant advantage of ULV application is that the biological performance in most cases is expanded for longer period. These two advantages have been achieved with ULV application of malathion. On the other hand all the insecticides are not suited for ULV application either due to their physical nature, high toxicity to man and animals or for lack of offering good biological performance. Some important desirable characteristics of an ULV product are :

1. It should have a very low volatility.
2. It should have a low viscosity.
3. Solvents (if used any) should have good solubility.
4. Solvent should be non-phytotoxic.
5. Solvent should be compatible with the pesticides.
6. The product should be safe for human being and animals.

On account of ideal characteristics for ULV application, malathion ULV concentrate is most widely used insecticide in various public health and agricultural programmes throughout the world. In India also, the usage of malathion ULV concentrate is well accepted. The recommendation of malathion ULV concentrate based on trials conducted in India; are given in the following table :

Crop	*Insect-Pests*	*Rate of Application*
Cotton	Jassids, thrips, whitefly, green weevil	360—480 ml/acre.
Paddy	Hispa, jassids, fulgorids	480 ml/acre.
Wheat	Flea beetle, pyrilla	480 ml/acre.
Sugar-cane	Pyrilla	480 ml/acre.
Vegetables	Thrips, Jassids, mealy bug, white fly epilachna beetle	360—480 ml/acre.
Jowar and other cereals	Deccan wing less grass-hopper, Ear head midge	480 ml/acre.

3

Crop Protection

Polyphagous Pests

LOCUST (Literal meaning of the word is 'Plague')

This pest is of International importance and the following three species are commonly found in India ; of which desert locust is predominant all over the country.

(a) Desert locust (*Schistocerca gregaria* Forsk.)

(b) Indian migratory locust (*Locusta migratoria* Linn.)

(c) Bombay locust (*Patanga succincta* Linn.)

Systematic Position

Order : Orthoptera

Family : Acrididae

Host plants : The locusts are the polyphagous pest and feed on all varieties of agricultural crops, forests, fodders etc. They do not feed on ak, neem, jamun and sesum.

Distribution : This noxious pest is widely distributed throughout the world, affecting various parts of the important continents, like Asia, Africa and Europe. In India the desert locust is found almost around the year in the desert areas of Rajasthan as well as in the sandy parts of Punjab and Gujarat.

Economic importance and damage : The locust swarms cause the collosal losses to a variety of vegetation even including the bark of trees. It is well known fact that in the past locust attack resulted in famines in several parts of the world. The outbreak of locust generally depends on the climatic conditions. A perusal of records have shown that locust outbreaks in India occurred in the year 1812, 1821, 1834, 1843, 1869, 1889-97, 1901-3, 1906-7, 1912-15, 1926-31, 1940-46, 1951-55, 1963, and 1993.

Life history : The desert locust passes through three stages in the course of its life cycle, *viz.*, egg, nymph and adult.

Egg : The eggs are laid in batches of 60 to 150 at a depth of 8 to 14 cm in moist sandy or loamy soil. After laying eggs, the female closes the mouth of the hole with a frothy liquid which ultimately hardens into a water proof plug. Generally, breeding starts with the onset of good monsoon showers. A female may lay about 500 eggs in her whole life span. The eggs are yellow in colour and resemble to rice grain. It is about 78 mm long and 1 mm girth. They hatch in 12-14 days in summer and 4 weeks during winter. At this stage a suitable amount of soil moisture is most essential for the proper development of eggs. If this prerequisite is provided by timely rainfall the eggs hatch into a vermiform larvae with membranous covering in which they are enveloped and they enter in the hopper stage of their life.

Nymph (Hopper) : The colour, size and habit of hoppers depend upon the phase of locust. Generally, there are two types of phases namely ; solitary and gregarious.

Solitary phase : In this type the hoppers are inactive and the individual nymph lives solitary or scattered. They are restricted in the desert breeding areas namely. Rajasthan, parts of Saurashtra and Baroda, Cutch, Hissar and Mahindargarh district in Punjab. These permanent desert homes of the locust undergoes certain changes in phase. The hoppers of solitary phase are of green colour and behave like a common grasshopper.

Gregarious phase : In this phase, the hoppers are very active gregarious, *i.e.,* the individuals tend to remain together. They breed rapidly and form swarms which leave the breeding areas and march in group to longer distances crossing the many countries. The hoppers of this phase are generally of black colour with pink markings when young, developing black and yellow markings as they grow old. The gregarious hoppers settle into bands, which sometimes extend over several square kilometres. They move in large numbers and devastate field after field thus leaving no green vegetation which comes across their way. As there are some individuals which are intermediate in habits, therefore they have been put into a separate phase known as transient phase.

The hoppers moult 5 times and after 3-4 weeks they changes into adult. During rainy season the main breeding grounds of desert locust in India are located in and around the desert of Rajasthan. With the end of monsoon, *i.e.,* by September the desert areas get drought affected accompanied by a rise of air temperature which generally the locust avoids. Therefore, the new hoppers leave the area and are carried partly eastward by the westernly winds into Punjab, U.P., Bihar and Bengal and sometimes southward into the peninsular areas including South India. Although these flights may found till December and January and most of them are destroyed by birds. Those which survive breed in spring in Punjab and adjoining areas of U.P. The swarms that are carried westward by north-easternly winds in the direction of Pakistan and Iran breed during winter rainfall in these areas. The next generation produced in these western areas during the spring months are carried eastward into the Indian desert areas by the westernly winds prevalent in May and June. With the advent of rainfall, the summer breeding occurs in the desert and thus breeding continues year after year. They generally fly at 18-24 km per hour and can travel long distances without resting.

Adults : The description of the adults of different species is given below :

Desert locust **:** The young ones are bright pink in colour but older one turn into bright yellow. They have 6-7 stripes on each eye. The legs are hard and head and legs are of yellow colour. The pink locusts are very active and are likely to cause great damage to crops.

Migratory locust **:** The adults of this species come from Pakistan in summer and generally mate in Rajasthan and Gujarat. The hind wings are of yellow colour and the stripes are absent in eyes. They are now under control and may be seen in small number here and there only.

Bombay locust **:** It is found in India, Cylone and Malaya. It is generally found in sporadic form and never in swarms. The hind wings are of pink colour and stripes are found in each eye. The adults of this species are found in quescent stage from October to July. In India, it is found in Mumbai, Gujarat and Tamil Nadu and only one generation is found in a year.

Control **:** The Indian Government has a permanent locust warning organization as part of the Central Directorate of Plant Protection, Quarantine and Storage. The staff of which is posted in the breeding areas for constant watching and receiving information regularly about locust population. The following methods may be used in controlling the swarms of locust :

1. The eggs can be destroyed by digging them. But this method is not economical.

2. The trenches may be dug at some distance in the way of marching hoppers and after falling in them may be burried alive.
3. The hoppers as well as adults may be killed by using flame throwers or with kerosene oil torches.
4. Poison bait may also be used to kill them, the same is spread around the bushes, where hoppers rest at night or on the way of matching hoppers in the day time. The following bait mixture has been found effective :

Wheat bran	—	95 parts
BHC dust	—	5 parts
Molasses	—	Little quantity

5. Dusting of 10% BHC at the rate of 20-25 kg/ha has been found effective and cheap.
6. Spraying the crop with 0.1% neem seed kernel powder was found very promising and it protects the crop for 3 weeks.

Termite

The termites are found practically all over the world and popularly known as white ants. There are two distinct categories of termites.

Subterranean : Their colony is always partly underground and starts by a pair of reproductive forms which enter the wood or earth at the time of swarming. The common Indian species belonging to this category are as follows :

(i) *Odontotermes obesus* Holmgren.

(ii) *Microtermes obesi* Holmgren.

Dry wood termites : Their colony is confined entirely to wood and starts by pair of reproductive forms which enter the wood above the ground at the time of swarming. The following are the important species :

(i) *Coptetennes fletcheri* Holmgren.

(ii) *Eutermes* sp.

Since *O. obesus* is the species mostly associated with the cultivated crops, hence its description is given here.

Systematic Position

Order : Isoptera

Family : Termitidae

Host plants : This is a polyphagous pest and feeds on variety of forest trees and agricultural crops. Among the crops grown wheat, maize sorghum groundnut and sugarcane etc. are severely damaged. The other species destroy wood in its various forms, *viz.*, wooden building structure, wooden furniture and all articles containing cellulose like paper even decaying matter like cow dung.

Distribution : The pest is prevalent in all the warmer parts of the world but commonly found in the tropical regions.

Economic importance and damage : The food of termite includes cellulose and they eat almost every thing which contains this substance. Though it is a polyphagous pest but mainly feeds on sugarcane and wheat. Termites attack both sugarcane stalks and setts. As result of their attack, the young canes even the shoots of older canes dry up and finally the growth is adversely affected. The cane sets are eaten up from the cut ends and are rendered hollow, their eye buds are damaged resulting the poor germination of crop. In unirrigated areas the young wheat crop is damaged badly, as they destroy the roots of seedlings are finally the plants are dried off completely. They destroy the crop at night and remain hidden in their nest during day time. The attacked plants can easily be distinguished by irregular cuttings and adhering to soil particles. The fruit trees are also attacked and on branches and stem the soil channels are formed under which they live.

Life history : Three developmental stages *viz.*, egg, nymph and adult are found. They are the social insects and live in colony. There is polymorphism in adult stage and king, queen, soldiers, workers are found. The members of the colony may be divided into (i) reproductive forms which include queen, king, complimentary forms (substitute caste) and colonizing forms (macropterous forms winged), (ii) sterile forms which include the workers and soldiers.

Egg : The winged reproductives come out in swarms from the nest generally in the beginning of monsoon. Swarming usually takes place in the day time and most of the individuals of the swarms are destroyed by birds etc. The survivals mate, shed their wings and burrow in the ground to form a new colony of which they become the king and queen. Their flight is known as nuptial flight. The queen lays first batch of 10-130 eggs about a week of swarming, but later on may be five years. Eggs are small kidney-shaped and yellowish in colour. The individual egg is 1/2 mm long and hatch out in a week or so depending upon the environmental conditions.

Nymph : The newly hatched nymphs are yellowish white in colour and about 1 mm long. They eat for some time the excreta of the king and queen and later search the food. Nymphs develop into different castes in 6-13 months after 4-10 moultings.

Adults

Reproductive Forms

Queen : The queen is the largest individual ranging 6.6-8 cm in length and 1 cm in thickness. This high dimension of the queen as compared to other members is mainly due to the enlargement of the abdomen which is full of eggs. She is the mother of the colony and lives in a specially prepared royal chamber which is situated in the centre of the nest at a depth of 1-6 feet below the ground surface. The queen is wingless, creamy white in colour and her abdomen is marked with transverse dark brown bands.

King : The winged male which remains with queen after nuptial flight is known as king. There is generally a single king in each colony which is smaller than queen and remains with her in royal chamber.

Complimentary forms : These are the apterous or brachypterous forms of both the sexes which maintain the numerical strength of colony in absence of macropterous forms. In a single colony their number may be in hundreds.

Colonizing forms : They are generally produced in rainy season and attracted by light. They are brownish in colour with two pairs of slender, dark brown, long narrowed wings which are used for nuptial flight and after which they are shed. These forms emerge in millions from the parent colony during monsoon and after finding their mate they crawl into the soil and start the new colony. They are thus the kings and the queens of future colonies.

Sterile Forms

Workers : Among the sterile castes the workers constitute the main labour force of the colony and represent 90% of the whole colony. They are about 6-8 mm long, dirty white in colour with brown head and eyes are small or absent. Their mouth parts are fairly strong and used as the working tools for all types of odd-jobs. They work for the whole colony by storing food, looking after the young ones and building up the termitarium.

Soliders : They are slightly bigger than the workers. Their mouth parts are well developed and used for very active defence and offence, particularly the mandibles are very large, projecting and shaped for a variety of uses. Their main function is to protect the colony but also help the workers in keeping the colony neat and clean by removing the dead and sickly members. Both workers and soldiers contain sterile individuals of both the sexes.

Control

1. The removal of dead or decaying matter or dry stubbles from the field is useful because they attract the termite.
2. The use of partially decomposed manure should be avoided.
3. Irrigation water with crude oil emulsion may be used to avoid the termite attack.
4. Mixing of any one of the following insecticide before sowing in soil has been found very effective.
   (a) Endosulfan 4% dust @ 30 kg/hect.
   (b) Lindane 1.3% dust @ 25 kg/hect.
   (c) Heptachlor 3% @ dust 35 kg/hect.
5. In standing crops any one of the following insecticides may be used with irrigation water.
   (a) Chlorpyriphos 20 EC @ 3.75 lit/hect.
   (b) Gamma`BHC 20 EC @ 3.75 lit/hect.
   (c) Heptachlor 20 EC @ 3.75 lit/hect.
6. Cane setts before planting can also be dipped in solution of 1% DDT or Chlordane or 0.25% BHC suspension.
7. The seed treatment of wheat with chlorophyriphos 20 EC has also been very effective, for this, use 400 ml per quintel of seed. Before application, the insecticide should be diluted with 5 litres of water and the emulsion sprayed over the seed. The seed should be turned over to ensure proper mixing and the treated seed should be left over night for drying before sowing.

Hairy Caterpillars

A number of species of hairy caterpillars occur in India which have dense tufts of hair on their body. They are the young ones of generally moderate sized moths which have often conspicuous colouration and are nocturnal in habit. A common characteristic of this group is that the moths as well as the larvae appear in large numbers during certain fixed seasons when climatic conditions are best suited to them. There are about 10 species of hairy caterpillars found in India, out of which following are the common species.

Family-Arctiidae

(i) *Amsacta moorei* Butler-Red hairy caterpillar

(ii) *Spilosoma obliqua* Walker-Bihar hairy caterpillar

(iii) *Utetheisa pulchella* Linn. Sunnhemp caterpillar

(iv) *Pericallia ricini* Fabar. Black hairy caterpillar

Family-Lymantriidae

(i) *Euproctis lunata* (Walker)

(ii) *Euproctis fraterna* (Moore) Castor hairy caterpillar

(iii) *Euproctis seintillus* (Wlk)

(iv) *Lymantria dispar*

Among this whole group of insects, the red hairy caterpillar and Bihar hairy caterpillar are the most injurious to agriculture throughout India.

Bihar Hairy Caterpillar

Scientific name: *Diacrisia (Spilosoma) obliqua* walker

Order : Lepidoptera

Family : Arctiidae

Host plants : It is a polyphagous insect which generally feeds on castor, til, mung, linseed, toria, mustard, some vegetable crops and medicinal plants.

Distribution : It is distributed throughout India and mostly in states of Bihar, M.P., U.P., Punjab and Rajasthan.

Nature of damage : The damage is caused by the larvae which are provided with biting and cutting type of mouth parts. The adults are practically harmless but responsible for progeny build up. The caterpillars in early stages feed gregariously on the lower surface of leaves and skeletonize them, *i.e.*, leaving their veins only. After a week, when they are grown up disperse over the field and eat the leaves and top shoots of plants. In case of heavy infestation the crops are completely defoliated. It has been observed that crops grown in rainfed areas having light soil receive more frequent endemic and epidemic of this pest.

Life history : The pest passes through four stages, *i.e.*, egg-larva-pupa and adult in its life cycle.

Egg : The moths of hairy caterpillars appear in large numbers after the first few good showers of monsoon and are attracted to light where they can be seen and easily distinguished. These moths emerge from the hibernating-aestivating pupae in the soil. After copulation, the female moth lays eggs at night on the lower surface of leaves. They lay eggs in large prominent batches and each batch may have 50-100 eggs. Single female can lay more than 1000 eggs in its life span of three to five days. The egg mases look-like clusters of poppy seeds specially conspicuous against the green foliar back ground. The eggs hatch within 4 to 5 days.

Larva : The larvae after hatching begin to feed on the lower epidermis of leaves and skeletonize them. They remain congregated for a few days then disperse and spread over field. The freshly hatched larvae measure 5-8 mm in length and possess dense crop of hairs on their body. The feeding propensity increases with their age. It moults six times to reach maturity and measures 40-45 mm in length. It has an orange coloured body with both end black and whole body is covered with hairs. The larval period lasts for 15-20 depending upon the environmental conditions.

Pupa : The full grown caterpillar sheds its hairs and pupates in the soil inside a rough cocoon mainly composed of hairs. The pupa is exarate, brown in colour and about 20 mm in length. After a lapse of 8-15 days the moths emerge out of these pupae during active period of its life cycle.

Adult : The moth is of medium size and pale buff in colour. There are black spots on the wing and on few last segments of abdomen. The eyes and antennae are black in colour. The moth measures 25-30 mm in length with wing expanse of 45 to 50 mm. The longevity of adults range from 5-7 days.

The life cycle is completed in 29 to 43 days depending upon the prevailing conditions. There are about 5-6 generations in a year. The insect hibernates during December-February and aestivates from May to June in pupal stage. The adults emerge with the onset of monsoon and thus the life cycle is repeated.

Control : As given in red hairy caterpillar.

Red Hairy Caterpillar

Hindi name : *Kutra, Kamla, Lal bhurli, Katra*

Scientific name : *Amsacta moorei Butler*

Order : Lepidoptera

Family : Arctidae

Host plants : It is a serious and polyphagous pest which generally feeds on jowar, bajra, maize, urd, mung, til, groundnut cowpea, cotton, guar and sunnhemp.

Distribution : This species is distributed throughout

Northern India and the another species *A. albistriga mung* is found in southern parts of India.

Nature of damage : The caterpillar is damaging stage which has chewing and biting type of mouth parts. Among the whole group of hairy caterpillars this is the most injurious to agriculture. The newly hatched larvae prefer to feed on the growing parts of plants gregariously and after few days they disperse in the whole field. When the vegetation of the field is finished they march out in large bands to neighbouring fields and continue to migrate from field to field. In years of severe outbreaks, fields may have to be resown in the beginning of the season and later on it becomes too late even for resowing. In July, 68 they appeared in epidemic form in eastern parts of U.P. covering more than 14,000 hectares of land. Authors experience has shown that out breaks of this are generally associated with the seedling stage of the crop. Further, their epidemics are positively co-related with the intermittent shower of rains followed by bountiful supply of food material.

Life history : Four developmental stages *viz.*, egg, larva, pupa and adult are found in its life cycle.

Egg : Female moth starts laying eggs after 6 to 9 hours of mating. The eggs are generally laid preferably on the under surface of leaves. They are creamy in colour and laid in prominent clusters. These egg masses look like clusters of poppy seeds specially conspicuous against the green back ground. An egg is 0.65 mm long, oval and dark brown in colour at the time of hatching. A female moth lays as many as 100-150 eggs during its life time. The incubation period (hatching time) lasts for 3-5 days and generally occurs early in the morning or late at night.

Larva : The dark coloured tiny caterpillars begin to feed immediately after hatching. They remain congregated for a few days and then disperse. Newly hatched caterpillar measures 1.5 to 2 mm long and 0.75 mm wide. It contains two black spots on dorsal side of 1st abdominal segment. The larva moults 6-7 times and becomes full grown which measures 40-50 mm in length and 6-5 mm in breadth. They are dark brown or deep orange in colour covered with reddish brown tufts of dense hairs. At the posterior and anterior ends of some larvae, a band of black hairs is also found. The larval period lasts for 15-23 days depending upon the environmental conditions.

Pupa : Generally, the full grown caterpillars begin to go down into the soil for pupation at the end of September. Before pupation a large number of larvae migrate from cultivated fields to the neighbouring uncultivated area. Here they pupate in the soil and remain there till the onset of monsoon rain in the following year. Pupation takes place in the soil in a cocoon made from larval hairs. The pupal colour is light brown which turns to dark brown after some time.

The pupa is 16-17 mm long and 7-8 mm wide. A timely rain on the eve of pupation is a pre-requisite for successful pupation for the caterpillars which require a loose wet soil for burrowing to a depth of 15 cm. The failure of rains at the critical stage will result in whole sale annihilation of the pest. The pupa remains in the soil for about 9-10 months generally from October to July. During active period of life cycle pupal period varies 7-12 days.

Adult : The adult is of medium size moth with white wings and black bands and spots on the body. The female is larger than the male. Female is 25 mm long with wing expansion of 42 mm while male is 17 mm long with wing expansion of 34 mm. The fore wings have red stripe at the margin of costal region and there are black spots on the hind wings. The moths are nocturnal in habit and positively phototrophic in nature. They survive only for 3-5 days.

The moths start laying eggs after one or two days of emergence. The life cycle is thus completed within 28-41 days. Generally, 1-2 generations are found in a year.

Control

1. Since the moths are attracted to light, the light traps should be set-up as soon as the rains start.
2. Collect the egg masses and young caterpillars in the early stage and destroy them.
3. The chemical control operation should be resorted as quickly as possible, *i.e.*, before larvae develop the migratory tendency. The following dusts have been found effective :

   (a) BHC 10% dust @ 40 kg per hectare.

   (b) Methyl parathion dust @ 25 kg/hectare.
4. The following emulsifiable concentrates are found quite effective when used at the rates given per hectare :

   (a) Endosulfan 35 EC-1.875 lit.

   (b) D D V P 100 EC-0.599 lit.

   (c) Phosphamidon 85 EC-0.500 lit.

   (d) Chlorpyriphos 20 EC-1.5 lit.
5. Spraying of neem seed kernel powder suspension (1%) makes the larvae unable to feed on the crop.
6. Thuricide (formulation of *Bacillus thuringiensis, Bert)* spray has also proved very effective in killing the larvae of this pest.
7. Following micro-organisms are the effective parasites of this insect :

   (a) *Aspergillus fumigatus Frs.* (Fungi).

   (b) Nuclear polyhedrosis and pox-like-virus.

White Grub

Hindi name : Safed gidar

Scientific name : *Holotrichia consanguinea Blanch. H. serrata Fabr.*

Anomala bengalensis Blanch.

Order : Coleoptera

Family : Melolonthidae

Host plants : White grubs are polyphagous pest and feed on almost all the kharif crops like jowar, bajra, maize, groundnut, sesamum, sunflower, chillies, cotton, sugarcane, tobacco, brinjal, cucurbit and bhindi, etc.

Distribution : They are of world wide importance as noxious crop pest and cosmopolitan in occurrence. White grub has attained a status of major devasting pest of a variety of rainy season crops in India, particularly in states like Rajasthan, Gujarat, Maharashtra, Karnataka, Tamil Nadu, Bihar, U.P. and Orissa.

Nature of damage : The damage caused by white grubs has been estimated to the tune of 40-80%. At times, the damage caused by grubs is so much so that entire stand of the crop is destroyed thereby necessitating the resowing in the field. During the last few years, white grub has posed such a serious and alarming situation in the country that it has been designated as a National pest.

The losses are inflicted by grub and the battle, the former feeds on fine rootlets, nodules in case of leguminous crops and then girdles of the main root, whereas the latter (adult) feeds on shrubs and the tress growing nearby the cultivated fields. The plants damaged by the grub gives a wilted appearance and finally dries out, while in case of beetles the attacked plants get defoliated. In case of severe grub infestation the entire crop is destroyed.

Life history : The life cycle of this pest passes through the following four stages, *viz.*, egg, grub, pupa and adult.

Egg : The beetles emerged from the soil after dusk (7.30-8.00 p.m.) followed by good premonsoon rains, which may occur from April to June. The emergence, movement and distribution of adults are mainly governed by rainfall, temperature and atmospheric humidity. Generally mating takes place, soon after the emergence of adults. After feeding and mating, the beetles get back to the soil in early morning hours and lay eggs. The eggs are generally laid singly in loose soil or in an earthen cell inside the soil upto the depth of 10 cm. The eggs are oval, creamy white when fresh and later turn to brown in colour. The mean number of eggs laid by single female ranges from 30-120. The hatching period varied from 7-14 days depending upon the prevailing conditions.

Grub : The newly hatched grub is creamy white and measures 10-12 mm in length and 2 mm in width. They feed on organic matter. There are three larval instars of the pest and the mean body length and width are 11.73 mm and 2.91 mm for the first instar, 17.88 mm and 4.73 mm for the second instar and 32.23 mm and 7.71 mm, respectively of the third instar. The full grown grub is curved C-shaped and dirty white in colour, its posterior abdomen is shiny black devoid of folds through which the intestinal contents are visible. The head is dark brown with strong mandibles and prominent thoracic legs. The larval period for the first and second instar ranges from 9.12 days and 3.4 weeks, respectively while in case of third instar it lasts for 6 weeks. Thus total larval period varies form 72 to 82 days.

Pupa : The full grown grubs move down deeper in the soil in search of moisture and for pupation. The grub constructs an earthen cell in which it passes a quiscent or prepupal stage which lasts for one to 6 weeks. The freshly formed pupa is light yellow and extremely tender, but as it grows older it turns brown and becomes some what hardened. It is of exarate type, *i.e.*, the appendages are free from the body. The pupal duration ranges from 10-27 days depending upon the type of soil.

Adult : The freshly formed beetle is cream coloured with

soft white elytra with the lapse of time the colour changes to brown and elytra gets hardend. Generally the adults are lamellate and males being smaller then females. These newly emerged beetles remain within the earthen cell and do not come out of soil until the onset of summer rains.

The total life cycle is completed from 90 to 108 days and all the known species in India have one generation in a year.

Control

1. The use of light traps is the successful method for the collection of beetles during emergence in the night.
2. The repeated ploughings, preferably soon after the summer rains, help in exposing the various stages of white grubs to their natural enemies.
3. The adult beetles can successfully be controlled by treating the trees with the following insecticides :
   (a) Carbaryl 0.1%
   (b) Monocrotophos 0.05%
   (c) Chlorpyriphos 0.05%
4. The grubs may be controlled by the soil application of following insecticides :
   (a) Phorate 10 g. @ 25 kg/hect.
   (b) Ekalux 5 g. @ 25 kg/hect.
   (c) Carbofuran 3 g. @ 35 kg/hect.
   (d) Oftanal 10 g. @ 25 kg/hect.
5. The sprinkling of chlorphyriphos 20 EC @ 5 litres per hectare offers the economic and effective control of grubs.

   The following micro-organisms are found parasisting the grubs :

   (a) Bacillus thuringiensis *(bacteria)*

(b) Bacillus popillae *(bact)*

(c) Beuveria bassiana *(fungus)*

(d) Aspergillus parasiticus *(fungus)*

6. Seed treatment of groundnut kernels with chlorpyriphos @ 15-20 ml/kg of seed has been found very effective in protecting groundnut crop against white grub.

Sugarcane White Fly

Local name : Lahi, Safed Mashi

Scientific name : Aleurolobus barodensis Mask.

Order : Hemiptera

Sub-order : Homoptera

Family : Aleurodidae

Host plants : Sugarcane, jowar, bajra, maize, wheat, barely etc.

Distribution : The pest is predominantly found in the state of Punjab, U.P., M.P., Bengal and Bihar.

Nature of damage : Nymphs and the adults of this pest cause damage by draining the sap from the leaf tissues. The attacked crop looks yellowish and pinkish from a distance. In cases of severe infestation, the leaves turn black due to the deposition of fungus on the honey dew excreted by the pest. As a result of this cane tops become unsuitable even for the cattle feed.

The juice quality of attacked canes is adversely affected. There may be a loss of 30-40% in sucrose and 20-40% in total solids. In such cases it is difficult to prepare good quality gur and to get a better sugar recovery in the mills. Generally, it has been observed that ratoon crops and the low lying fields having more moisture or stagnation of water receives more infestation of this pest on sugarcane crop. Besides, in want of nitrogen the incidence of this pest is increased.

Life history : This pest exceptionally passes through four stages, *i.e.*, egg, nymph, pupa and adult to complete the life cycle. The description of these stages is as follows :

Egg : The eggs are laid on lower surface of leaves in small straight rows, parallel to mid rib having about 51 eggs in each row. A female lays eggs at different intervals in her life time and on each occasion it lays about 100 eggs. The egg is cone shaped and provided with a tiny little pedicle by which it is attached to the leaf. Each egg measures 3.3 mm in length and looks pale yellow in colour when freshly laid. Later, it changes to orange to grey dark grey and finally to black with a brownish tinge. Egg stage lasts for 7 days is summer and 13-14 days in winter.

Nymph : The nymphs, after hatching, settle down on the lower surface of leaves. They are shining yellow in colour and measure 0.31 mm in length. Finally, their colour change into black. They continuously de-sap the plant and ultimately render it pale and sickly in appearance. Each leaf may bear several hundreds of small black nymphs on its lower surface. There are three moults; before 1st moult the nymphs have 4 segmented antennae and 3 pairs of fully developed legs whereas after moulting the legs become vestigeal and antinnae disappear altogether. At this stage they are found stiking with leaves and sucking their sap. After second moult the nymph measures 2 mm in length and having vasiform orifice at its terminal end wherein the female reproductive organs are situated. The honey dew is excreted from the anus. The nymphal period lasts for 11 days in summers and 31 days in winters.

Pupa : This is the inactive stage of the pest which is covered by the white waxy powder. The pupa is flat, oval, greyish in colour and bigger in size than nymph. There is a 'T' shaped white marking on the thoracic region from where the pupa is splitted. Generally, pupal period is 7 days which prolongs in winters.

Adult : Emergence of adults usually takes place in the morning hours. They are tiny little creatures with pale yellow body and mealy wings. The male is comparatively smaller and sluggish than the female. Soon after emergence, males fly to the females for mating. Usually egg laying starts within half an hour after mating. The adults live for about 30 days.

White fly usually becomes active with the start of monsoon in July. High humidity during the monsoon period provides favourable condition for its multiplication, therefore, its maximum attack is recorded during September-October. Later, in winters its activity slackens and again becomes active in springs, *i.e.,* during March-April. Hot and dry winds which follow in May and June do not allow it to multiply till the conditions again become favourable in July. The pest completes its life cycle in 27 days during summer which, however, extends in winters.

Control

1. The crop should be properly looked after and balancely manured particularly with nitrogenous fertilizer. In low lying areas where water logging is common, ratooning of the crop should be avoided.
2. The resistant varieties of sugarcane *viz.,* Co 285, Co S-385 and 1158 should be grown in affected areas.
3. Spraying the crop with following insecticides have been found quite effective @ 1000-1200 lit/hect.

   (a) DDT 50 W.P. -5.0 kg.

   (b) Endosulfan 35 EC-1.0 lit.

   (c) Formothion 25 EC-1.0 lit.

   In cases of severe infestation two sprayings are generally

recommended; one in September and the other in October.

4. The following insects are the natural enemies at different stages of this pest :

   (a) *Azotus delhiensis* : on pupae

   (b) *Amitus aelurolobi* : on nymphs and pupae

   (c) *Exetmcoeras delhiensis* : on nymph

   (d) *Encarsia sp.* : nymph and pupa

Sugarcane Leaf Hopper

Local names : Alghora, Pyrilla

Scientific name : *Pyrilla perpusilla* Walker

Order : Hemiptera

Suborder : Homoptera

Family : Fulgoridae

Host plants : Sugarcane, wheat, jowar, bajra, barley and other plants of the graminae family.

Distribution : Generally, it is found in all the sugarcane growing areas of the country, but it is more prevalent in U.P., Maharashtra and M.P.

Nature of damage : Both nymphs and adults suck the cell sap from under surface of the leaves and thus the attacked leaves become pale and wilted. The honey dew excreted by the pest attracts shooty mould which interferes with the proper functioning of leaves and thus renders them unfit even for the cattle feed. In a grown up crop attack of the Pyrilla causes deterioration in juice quality as a result of which sugar recovery is adversely affected and the quality and yield of gur also suffer when infestation is severe, the recovery of sugar may fall to as low as 75 per cent of the normal. In areas of severe infestation it becomes difficult to prepare good quality of gur.

Kalra (1973) reported that the sucrose percentage in juice is reduced by 2-5 units. Ordinarily the recovery of gur should be 10 per cent but is generally less by 2.2 to 4.5 per cent from pyrilla infested crops.

The pest is active through out the year, but more damage is caused from May onwards and it becomes worst in October and November after the monsoon. It has been observed that moderate to high humidity (60-80 per cent) coupled with low maximum temperature (35 to 38) during summer months and low rainfall or long dry intervals during the monsoons are associated with the occurrence of the pest in a serious form. The ratoon fields with high level of nitrogen are generally infested more in comparison to normal fields.

Life cycle : There are three stages in its life cycle *viz.*, egg, nymph, adult.

Egg : The female after copulation lays eggs in clusters of 2-60 generally on the underside of the leaves; in grown-up crop the eggs may be laid on the lower dry leaf sheaths also near the midrib. The eggs are covered with white cottony filaments from the anal tuft of the female and are bright green yellow or dark brown in colour, oval in shape and 1mm in diameter. The eggs hatch in 7 to 12 days during summer and 3 to 4 weeks during October-November. In winter the eggs are deposited on the leaf sheath, so that there may be no effect of cold. Generally, eggs are laid from April to November and during severe infestation they may be laid down on grasses, wild plants and even on ground.

Nymph : The nymphs which hatch out from the eggs are pale brown in colour and 1.3 mm long. At this stage they do not possess snout, wings and tufts. Within a week they are characterised by the possession of two long tufts of white waxy secretion at the end of the abdomen. They start sucking the sap of the canes, grow in size and change into adults after 5-6 instars usually, there are 5 nymphal instars. The total duration of nymphal stage may vary from 4-6 weeks in summer and 15 to 22 weeks during winter months.

Adult : The adult hopper is straw coloured measuring 9 mm long with wing expansion of 23 mm. There are two pairs of wings folded roof shaped on the head extended anterior from a beak like structure snout. The head is provided with white black large eyes and 3 segmented antennae the females can be distinguished from the males by the presence of pair of pad on the posterior end of the body. Prominent bundles of white waxy threads develop over these pads during the egg laying period. The female survives 6 to 8 weeks, while male only 4 to 6 weeks.

The life cycle from egg laying to adult stage is generally completed in 42 to 63 days and there are 4 generations in a year. The first 2 broods are clear-cut but the third and fourth overlap each other. The eggs of Ist generation are laid in first week of April, IInd generation from late June to middle of July, IIIrd generation in the middle of August and fourth generation are laid during November and December after which adults usually die due to cold. The cycle is continued through the nymphs which develop in to adults in the following March, thus completing four generations during the year.

Control

1. Burn all trash after harvesting the sugarcane.
2. Remove all the leaves bearing egg-clusters during March-April.
3. Ratooning should be avoided in badly infested areas and balance N_2 should be given.
4. The resistant varieties like Co 386 and Co 975 may be grown, if possible.
5. The following dusts have been found very effective in controlling the pest :

(a) Malathion 5% @ 40 kg/hect.

(b) Carbaryl 10% @ 30 kg/hect.

6. The crop may also be sprayed by any one of the following spray to protect the infestation of pyrilla :

    (a) Endosulfan 35 EC @ 1.0-1.5 lit in 1200 lit of water/ hect.

    (b) Phosphamidon 100 EC@ 250 ml.

    (c) Dimethoate 30 EC @ 0.750 to 1.00 lit.

    (d) Malathion 50 EC @ 1.50 to 2.00 lit.

    (e) Malathain 96% LVC (ULV spraying) 1 to 1.25 lit.

7. *Natural enemies* : The following are the effective natural enemies of the pest :

    (a) Egg parasites :

    - (a) *Tetrasticus pyrillae*
    - (b) *Aeginiaspis pyrillae*
    - (c) *Cheiloneurus pyrillae*
    - (d) *Oonocystus pyrillae*

    (b) *Egg predator* : Brumus saturalis

    (c) *Nymphal parasites :*

    - (a) *Epipyrops melanoleuca*
    - (b) *Lestodryinus pyrillae*

    (d) *Fungal Parasites* : Metarhizium anisopliae.

Sugarcane Top Borer

Local name : Choti bedhak, Salai, Gabsukh and Funga.

Scientific name : *Tryporyza (Scirpophaga)* nivella Fabr.

Order : Lepidoptera

Family : Pyralididae

Host plants : Sugarcane, Jowar, Kans *(Saccharum spontaneum)*, Sarkanda *(Saccharum sara)*. and Dhab *(Eragresits csnosureid)*.

Distribution : All sugarcane growing areas specially in U.P., M.P., Bihar, Punjab, Chennai and Mysore.

Nature of damage : The damage is caused by the caterpillar which bears chewing and cutting type of mouth parts. Its attack starts on the growing crop from the month of May. The newly hatched larvae move on the leaf surface for 3 to 4 hr and then after selecting a suitable spot on the first open leaf on the top, penetrates in to the mid-rib from the underside of the leaf and bores its way towards the axil. It comes out to the mid-rib and bits in to whorl of newly developed leaves which cause the series of holes and may be seen when the leaves open. Thus, it goes to the spindle and moves down towards the growing point and destroy the top portion of the plant. The central shoot when tunneled and damaged by the larva, dries up and resulting in the formation of dead-heart. When the central shoot is damaged, side shoots develop and produce a sprout and branched top known as bunchy top.

Generally, it is found through out the season but major damage is caused from July to September. The loss is caused to the tune of 30 per cent. The bored cane mature early but deteriorate from January onwards when a decrease of one or two units in sucrose per cent in juice is observed.

Life cycle : Four developmental stages are found in its life cycle *viz.*, egg, larva, pupa and adult.

Egg : The female moth lays eggs in clusters of 5-8 mostly on the under side of the leaf which is covered with brown hairs from the anal tuft. Eggs are flattened, scale like and overlap one another in the cluster. A single female can lay about 500 eggs. They are round, flat and orange brown in colour measuring 1.2 mm long and 1.0 mm wide. At the time of hatching they become light brown in colour, incubation period lasts 9-12 days.

Larva : The newly hatched larva is white in colour with black head and measuring 22 mm long. After remaining on the leave it enters the top shoot of the cane through the mid-rib of the leaf grow in size and goes on penetrating the top till 4 to 6 internodes are bored. The full grown larva measures 28 mm long, 45 mm thick and creamy white to light brown in colour, It has a dirty stripe on the back. The caterpillar constructs a silken gallery for pupation and exit hole for emergence of the moth. The period taken by the caterpillar in attaining the full growth is about 3 weeks in summer and rainy season and 5 weeks in winter. It remains in hibernation during winter and stimulates activity in February.

Pupa : Full grown larva pupates inside the cane having mouth towards exit hole. The pupa is shining brownish yellow in colour measuring 16-18 mm long. The pupal stage lasts 10-12 days and out of which adult moth emerge out from exit hole.

Adult : The adult moth is silvery white, 2.5 mm long and with a wing expanse of 3 mm. The abdomen of the male is slender and pointed while that of female is stout and covered with crimson or orange hairy anal tuft. The antennae are filiform and tarsi segmented. The fore wings are long, narrow and bear hairs on the outer and posterior margin. The moths are nocturnal in habit.

The life cycle is completed from 42 to 51 days and there are 6 generations in U.P. while only five in Bihar and Punjab. The pest over-winter in caterpillar stage and regain activity in the beginning of February. The eggs of Ist generation are laid in March on ratoon sprouts or autumn planted crop, while of second generation in the month of May. The attack of IInd generation is generally mild as conditions (hot weather) prevailing at that time are not suitable for its activity. The attack of the IIIrd generation starts in July and of fourth in the middle of August. The sixth generation attack appears in the end of September or early October and caterpillars of this generation enter into hibernatrion within the cane.

Control

1. The moths may be destroyed by attracting them in light traps.
2. Adequate and timely nitrogen holds the crop to withstand the attack of top borer.
3. The egg masses may be destroyed easily as they can be spotted out due to reddish-brown hairy mass covering them.
4. A campaign may be organised to remove the top bored plants in the month of July.
5. Ratooning should be avoided where the attack of the pest is usually serious and water-logging condition should also be avoided specially in rainy season.
6. Grow following resistant varieties, e.g., For U.P. Cos 761, 796

   For Bihar B.O. 10, 22 and 24

   The following varieties have been found better to different parts of U.P.

   Eastern-Co. 395, 453

   Western-Co. 186, 243, 421 and 443

   Central & Ruhelkhand-Co 313, 421 and 453.
7. Sprinkling of endosulfan or carbofuran granules over the top portion of the crop @ 5 and 15 kg per hectare, respectively were found quite effective.
8. Spraying the crop with 0.1% endosulfar (35 EC) @ 1200-1500 lit/hec minimises the attack of borer.
9. Recent studies have shown that soil application of

carbofuran or phorate granules @ 2 kg a.i. per hect. is also very effective in controlling the pest.

Natural enemies : The following are the important natural enemies which keep the pest population low up to certain extent :

Egg parasite : *Trichogramrna sp.*

Larval parasite : *Apanteles flavipes.*

Pupal parasite : *Stenobracon deesae.*

Sugarcane Shoot Borer

Local name : Ankur bedhak or Stambh bedhak

Scientific name : *Chilo* (Chilotraea) *infuscatellus* Snell.

Order : Lepidoptera

Family : Pyralididae

Host plants : Sugarcane, Maize, Jowar and Bajra etc.

Distribution : Besides India, it is found in Mayanmar, Japan and Korea. In India it is mainly distributed in U.P., M.P., Punjab, Bihar and Maharashtra.

Nature of damage : The damage is caused by the caterpillar which bore the cane stalk at the level of the ground and feed inside the soft tissues, going upwards as well as down wards as far as the roots. The central whorl of leaves dries and forms characteristic dead-heart. The peculiarity of the dead-heart is that the only central leaf dries up and other leaves remain green for a long time. When dead heart is pulled out, it gives very offensive smell. The shoot borer caterpillar is migratory in habit and, after killing one shoot it goes to another and a single caterpillar may destroy three or four shoots. The long tunnelling of stem greatly reduces the yield and quality of the juice. Under severe infestation it may cause 25-30 per cent loss to the crop. This pest is more serious in western and central cane growing areas of the Uttar Pradesh. In Northern India, maximum damage is caused during April to June. The incidence of shoot borer is more pronounced in light soils under dry conditions.

Life cycle : Four developmental stages namely egg, larva, pupa and adult are found in its life cycle which are described below :

Egg : The eggs are laid in clusters by female moth on the under surface of the leaves. They are creamy white, oval, scale like and deposited in two or three overlapping rows. There are more than 30 eggs in each cluster. The eggs hatch in about 4 to 6 days.

Larva : Newly hatched larva possesses a dark head and translucent body with long hairs on dark spots. These spots change into prominent grey tubercle provided with long bristles and five stripes of violet colour develop. The newly hatched larvae feed singly or in groups on the tender leaf sheath for 4 to 6 days. The behaviour at this stage is more or less like a leaf miner. Later, they disperse and each caterpillar makes a hole on the stem just at the soil surface. After entering in to the cane stalk they feed on plant tissues for 3 to 4 weeks. The full grown caterpillar is 25 mm long and of dirty white in colour. Larval period lasts 3 to 4 weeks and then it changes into pupa.

Pupa : Full grown larva pupates within the tunnel in the shoot and before pupation it constructs an exit hole in the cane just above the ground surface for the emergence of moth. The pupa is brownish in colour. The pupal period lasts for a week and thereafter a moth emerges through a hole in the stem. During winter larvae do not change in to pupae but hibernate and pupate in the month of March.

Adult : The adult moth is greyish brown in colour with dark markings on the fore wings while hind wings are greyish white.

Eyes large, antennae filiform with a wing expanse of 25 to 35 mm. The posterior region of the abdomen is some what pointed.

The adult moths after emergence copulate in the night and females start egg laying after 2-3 days of copulation. The life cycle is completed in 34-44 days. There are generally 5 to 6 generations in a year but during favourable conditions 8 generations in a year have been recorded. Pupation among the over wintering larvae starts in the later half of February or beginning of March when there is slight rise in temperature.

Control

1. The stubbles may be destroyed and burnt after harvest.
2. Earthing of sugarcane should be done during May and June and the surface portion of cane may be covered.
3. Timely cutting of attacked shoots at or just below the ground level ensures destruction of most of the caterpillars present within the stem.
4. Larvae may also be destroyed by pulling the dead hearts.
5. Early planting (by middle of February) is helpful in checking serious damage of shoot-borer under North Indian conditions.
6. Treat the soil with 10% BHC dust @ 30 kg/hec before sowing.
7. Sprinkling of gamma BHC, Heptachlor or Telodrin emulsion at 1 kg a.i. per hec mixed with about 1,500 litres of water over the sets in the furrows at the time of planting of cane proved effective.
8. The spraying of endosuflan 35 E.C. @ 1.85 to 2.50 lit in 1200 lit of water was found effective. This should be repeated 3 to 4 times at 10-12 days intervals.

Natural enemies : The following natural enemies have been found on different stages of the pest :

Egg parasite

(a) *Phanurus beneficiens*

(b) *Trichogramma minutum*

(c) *T. nana*

Larval parasite

(a) *Apanteles flavipes*

(b) *Stenobrancon nicevillei*

(c) *Goniojus indicus*

Pupal parasite

(a) *Tetrasticus ayyeri*

(b) *Aperostocus sp.*

(c) *Melcha ronetipinis*

Gurdaspur Borer

Local name : Gurdaspur, bedhak, Dehradun bedhak

Scientific name : *Acigona (Bissetia) Steniellus Hmpsn.*

Order : Lepidoptera

Family : Pyralididae

Host plants : Sugarcane, Jowar and Maize etc.

Distribution : The pest was originally recorded from Sialkot (Pakistan) and Gurdaspur district (Punjab) in the early twenties. Now it is found in all sugarcane growing areas specially in Punjab, Haryana, Uttar Pradesh, Rajasthan and Madhya Pradesh. In U.P. it's incidence was found in Saharanpur, Muzaffarnagar, Dehradun and Rampur districts.

Nature of damage : The attack of Gurdaspur borer starts with the commencement of monsoon rains in the later part of June or July. Damage is done by the caterpillar which is quite characteristic. The caterpillar after entering the internode, moves upwards in a spiral manner making minute holes on the bud from within the internode, with the result that canes become weak, dry up or fall down easily in strong wind or heavy rain. Later on it bores in to central portion of the cane and feeds in a single straight tunnel moving upwards.

At the beginning larvae feed gregariously but after 10 to 12 days they disperse to adjoining canes. As the attack proceeds, whorl of leaves at the top dries up due to which further growth ceases. The damage normally varies from 15-25 per cent of the total crop, but in years of severe attack it may be 40 to 50 per cent in certain fields. *Kapoor* (1957) estimated the loss caused by this pest up to the extent of 38-100 per cent.

Life cycle : There are four developmental stages of the pest namely egg, larva, pupa and adult.

Egg : The eggs are laid by female in clusters in and along the groove of the midrib on the upper (ventral) surface of the leaf near the leaf sheath. Newly opened leaves in the top portion of the shoots, are preferred for egg laying; usually each cluster contains 20 to 50 eggs. Initially they are creamy white in colour which subsequently change in to brown. The eggs are flat and 1 mm in diameter. The egg stage lasts for 2-5 days.

Larva : The newly hatched larva after hatching feeds for some time on the bud situated on the top first or second internode and then bores into it from near the bud. During this stage as many as 40 to 50 larvae may be found in an internode. The larvae are pinkish in colour measuring 2.5 mm long. They come out after a period of 7 to 10 days and disperse to surrounding canes, each caterpillar penetrates separate cane. The full grown larva is 30 to 35 mm long having 4 violet stripes on the back and pinkish brown in colour. The larval period is of 21 to 34 days.

Pupa : Pupation occurs inside the cane. The pupa is brown in colour measuring 10 mm long and 35 mm wide. The pupal period lasts 6-7 days after which moth comes out through the exit hole prepared by the caterpillar.

Adult : The moth is greyish-brown with dark markings on the fore wings and measuring 10 mm long and 3.5 mm wide. The hind wings are greyish white with wing expanse of 25 mm.

The moths copulate on 2nd day of emergence and female starts egg laying after 2 days. The total life cycle is completed in 31 to 48 days and there are 3 generations in a year. The pest remains active from July to November. From July to October all stages of the pest are seen simultaneously because of overlapping generations. After November the caterpillars remain in hibernation during winter and in quiscent stage during early summer within the sugarcane stubble and become active after the first shower of rain.

Control

1. The attacked shoots should be regularly removed and destroyed.
2. The rougeing should be done from July to September, when young caterpillars feed gregariously.
3. Resistant varieties should be preferred, *e.g.*, Co 1157, 1158 and Co 6812.
4. In endemic pocket, the setts of cane should be treated for 2 hours in 0.2% trichlorfon (dipterex) before planting.
5. The following insecticides have been found effective in minimising the incidence of pest in standing crop. These may be sprayed 4 times at 15 days intervals @ 1200 to 1500 litre of water :

   (a) Gamma B.H.C.3.75 lit.

   (b) Endosulfan 35 EC-2.0 to 2.5 lit.

Natural enemies : The following are the effective enemies in different stages of the pest.

Egg parasite : *Trichogramma sp.*

Larval parasite

(a) *Apanteles flavipes*

(b) *Struniopsis interens*

Pupal parasite : *Isotima javensis*

Sugarcane Root Borer

Local name : Mool bedhak, Eekh ke Jar Ki Surhi, Jarkata, Makora.

Scientific name : *Emmalocera depressella* Swinh.

Order : Lepidoptera

Family : Pyralididae

Host plants : Sugarcane, Maize.

Distribution : It is found in all sugarcane growing areas of the country specially in U.P., M.P., Punjab, Bihar and South India.

Nature of damage : The damage is caused by the caterpillar during early stage of the crop. Its damage starts during April and caterpillars of the pest bore at the base of the newly formed stem and feed inside. As a result of which main shoot is damaged and dead-heart is formed. The damage is very peculiar, because larvae enter at the base of the stem which is very close to the surface of the soil. Further, they do not move upwards but remain near the base of the shoot. The dead heart caused by the root borer resembles that of shoot borer but may be differentiated on the following points :

1. Only one hole is found on the shoot.
2. Absence of offensive smell in dead heart.
3. Alongwith central leaf sheath, few side leaves are also dried and involved in dead heart.

4. There is no sign of attack (eating holes) on the leaf sheath.

The attack of the pest is more pronounced in ratoon crops and it persists from April to the end of September or middle of October. If the plants are young they may die but the older ones dry up and fall down. During rainy season its attack is minimised. Due to the attack of this pest there may be a decrease in yield by about 10 per cent reduction in sucrose percentage in juice by about 0.3 unit. In cases of severe infestation it may cause a loss to the tune of 40 per cent.

Life cycle : Four developmental stages namely egg, larva, pupa and adult are found in its life cycle, which are described as under :

Egg : The female lays eggs after copulation in the night on the underside of the leaves of the sprouting canes during April and May and lower most part and even on the soil surface. The eggs are deposited either singly or a few of them in clusters. They are flat, oval, creamy white in colour and measuring 1.2 mm long and 1 mm wide. The hatching period is about 7 days.

Larva : The newly hatched larva is light yellow or white in colour with a brown head measuring about 2.00 mm long. The larva enters the plant near the roots by making a small hole and feeds inside, if plant dries out it migrates to the another plant. It moults 4 times and becomes full grown which is creamy-yellow in colour with brown head and measuring 30-35 mm long. The larval period lasts for 40-45 days and before changing in to pupa it constructs a silken tube for emergence.

Pupa : The pupa is reddish-brown in colour measuring 17-20 mm long and 4 to 5 mm wide. The pupal stage is not found during winter. Generally, the pupal period lasts in 7-10 days which may extend up to 15 days depending on the environmental conditions.

Adult **:** The adult moth is dirty brown in colour, measuring 17 to 20 mm long with the wing expanse of 27 mm. The hind wings are white, shorter and wider than the fore wings and have a dark longitudinal stripe on each wing. The head is small with large black eyes and bipectinate antennae in the case of male and setaceous in case of female.

The female starts egg laying after 2 days of emergence. The total life cycle is completed in 57 to 73 days and there are 4 generations in a year, however 5 generations have been found in Punjab. In U.P. the larvae of 4th generation remain in hibernation during winter. The over wintering larvae are revived to activity in April when moths of 1st brood emerge and start egg laying. The maximum activity has been found in the month of August.

Control

1. Timely cutting of infested shoots at or just below the ground level is found useful
2. After harvesting stubbles should be destroyed and ratooning be avoided in endemic areas.
3. The resistant varieties *viz.,* Co 313, 513, 1148 and 1158 should be preferred to grow.
4. Sprinkling of gamma BHC @ 6.05 lit per hectare mixed with about 1875 litres of water over the sets in furrows at the time of planting found effective.
5. At the time of planting dusting of 10% BHC dust @ 30-35 kg per hectare in furrows was also found effective.
6. The spraying of endosulfan @ 1.85 to 2.50 litres/hectare in 1200-1500 lit of water gave good control when applied 4 times in standing crop at an interval of 10-15 days.

Natural enemies **:** The following are the effective natural enemies of the pest :

Egg parasite

(a) *Trichogramma minutum*

(b) *Phanurus beneficiens*

Larval parasite

(a) *Stenobracon deesae*

(b) *Phanerotoma sp.*

(c) *Chelonus narayani*

(d) *Apanteles sp.*

Jowar and Maize Stem Borer

Local name : Mokora, Makaya, Obra, Ratwa, Kanka ghundi, Khera, Tana chhedak.

Scientific name : *Chilo partellus (zonellus)* Swinhoe

Order : Lepidoptera

Family : Pyralididae

Host plants : Jowar, Bajra, Maize and Sugarcane etc.

Distribution : It is found practically in all the tracts of India where these crops are grown. Its distribution extends in the region of the world from Afghanistan to Indonesia and Taiwan in the east and Sri Lanka in the south and it has also been reported from east Africa.

Nature of damage : The freshly hatched larvae begin to feed on the tender leaves, particularly in the central whorl; after feeding for few hours on leave, they enter through the mid whorl into the shoot. The caterpillar attacks all parts of the plant except the roots. In case of maize only young plants while in jowar all stages of plants are attacked. In case of younger plants the growing point and the base of the central whorl are infested resulting in the drying up of the central whorl and forming the dead heart. In case of older plants, the stem is riddled with.

The attacked plants remain stunted in growth and produce no grain. More than one caterpillars are found in a plant. Generally 20 to 30 percent damage is caused by the pest and it is serious in dry years. *Jotwani* (1971) estimated 83.7 per cent loss in grain yield. Maximum damage is caused in the month of August to maize while during September and October to jowar. The pest remains active in the field from March to November.

Life cycle : Four developmental stages *viz.*, egg, larva, pupa and adult are found in its life cycle.

Egg : The female moth starts egg laying after two days of copulation. The eggs are laid in clusters on the lower surface of the leaf near mid-rib which are scale like and yellowish creamy white in colour. Each cluster may have about 4 to 20 eggs and overlapping one another in disposition. An individual egg is 1 mm long and turns in to brown and black at the time of hatching. The hatching period varies from 2 to 5 days. A female lays 250-300 eggs in her life span.

Larva : The freshly hatched larva is white in colour and about 2 mm in length. The head and prothorax are black and body is provided with four long brown stripes on its back. They feed about a week on the leaves then enter in to the stem and remain inside whole life. Generally, larval period is of 21-28 days but it extends to 193 days in winter as the caterpillar remains in hibernation. The full grown larva is dirty white in colour and having black spots on the body which are provided with a brown hair. It measures 27 mm in length.

Pupa : Pupation takes place in specially constructed chamber in the stem. Pupa is reddish-brown in colour and the tip of its abdomen is furnished with six spines arranged in a group of three. The pupa is about 10 to 13 mm in length. The male pupa is narrow and smaller than that of female. The pupal period lasts in 4-7 days and after which the moth emerges from the stem through a hole made by the larva at the time of pupation.

Adult : The adult moth is medium sized insect with a wing expanse of 25 to 30 mm and yellowish brown in colour. The fore-wings are pale straw coloured while the hind-wings possess a double row of black spots along their outer margins. The tip of the abdomen in female is dilated and tufted with hairs, while in male it is pointed and devoid of tuft.

The adult moths copulate after emergence. The male dies after copulation while female after 2 to 3 days of egg laying. They are nocturnal in habit and during the day remain hidden under dried leaves, clods of earth etc. The life cycle is completed in 29 to 41 days and there are 4-5 generations in a year. The moths from over-wintering larvae appear in March-April and start egg laying.

Control

1. The stubbles should be collected and destroyed, when harvesting is over.
2. Collection and destruction of all the dead hearts and attracting the moth in light-trap proved effective to control the pest.
3. The following insecticides have been found effective when sprayed in standing crop at the interval of 2, 5 and 7 weeks of sowing.
   (a) Dimecron 100 EC-200, 250 and 300 ml/hect.
   (b) Endosulfan 35 EC-1.0, 1.25 and 1.5 lit/hect.
   (c) Carbaryl 50 W P-1.0, 1.5 and 2.0 kg/hect.
4. The placing of following granules in the central whorl of the plant has been found quite promising but this application should be followed by second granular application or spraying with any one the above insecticides after 3 weeks of placement:

(a) Lindane 6% granule-20 kg/hect.

(b) Endosulfan 4% granule-15 kg/hect.

Natural enemies : The following are the effective natural enemies :

Egg parasite

(a) *Trichogramma minutum*

(b) *Bracon chinensis*

Larval parasite

(a) *Microbracon chinensis*

(b) *Apanteles flavipes*

(c) *Xanthopimpla sp.*

(d) *Blepharipoda sp.*

Sorghum Shoot Fly

Scientific name : *Atherigona socata* Rond

Order : Diptera

Family : Anthomyidae

Host plants : Sorghum, maize, wheat, broom corn, small millets and grasses.

Distribution : It is widely distributed in Europe, Africa and Asia. In India the pest is more serious in Southern parts.

Nature of damage : The maggots cause damage to the seedlings as well as to the early stage of the crop when it is in the six leaf stage. After six weeks of sowing the crop is rarely attacked. The maggots feed on the main shoot with the result growing-point is destroyed. The death of central short causing a typical symptom called "dead heart". The damaged seedlings may be killed or it may produce few non-productive tillers. When the plants are attacked in some what older age, the tillers are produced which mature later than the main crop. It may cause loss to the crop as high as 60 per cent. Extremes of temperature as well as heavy continuous rains adversely affect the population of shoot fly.

Life history : The four developmental stages namely egg, larva, pupa and adult are found which have been described below.

Egg : The female lays whitish cigar-shaped eggs, measuring about 1.2 mm in length, singly on the underside of the leaves of seedlings. Generally, only one or two eggs are laid on each seedling. The fly prefers to lay eggs on the central leaves. The eggs are provided with two wing-like lateral projections. The incubation period of eggs varies from 1-2 days.

Larva : The newly hatched tiny maggots creep out and reach in between the sheath and the axis, and bore into the shoot. The maggot is dirty whitish in colour and develops inside the shoot by feeding on decaying tissues. The larval period lasts for 6-10 days. The full grown larva is yellow coloured and measures 5.5 to 5.6 mm in length.

Pupa : The full grown maggots pupate either inside the stem or come out and pupate in the soil. The pupal period lasts for 6-9 days. The colour of pupa changes from initial light brown to deep brown. Pupa is of object type and its anterior end is round and is fitted with circular disc bearing the two spiracles.

Adult : The adult shoot fly is about 5 mm in length and be identified by the presence of three pairs of black spots on pale grey abdomen. They are more active in the early mornings or during evenings. The adult usually survive 3-4 days.

The life cycle of fly is completed in 14-23 days and many generations are found in a year. In northern India, the pest over-winters in the pupal stage.

Control

1. High seed rate may be used.
2. Roguing of attached plants on the appearance of dead hearts.
3. The resistant varieties *viz.;* CSH-7 and CSH-8 may be grown to reduce the damage.
4. Application of phorate or disulfoton granules in the furrows at the five on sowing @ 1.5 kg a.i./hect.
5. Seed treatment with carbofuran at the rate of 5 parts insecticide in 100 parts of seed is also quite effective.
6. Sprayings of 0.025% methyl demeton or phosphamidon is also effective, this should be repeated after 3 days intervals.

Earhead Caterpillar

Scientific name : *Cryptoblabes gnidiella* Mab

Order : Lepidoptera

Family : Pyralididae

Host plants : Jowar, Ragi and Bajra.

Distribution : The pest is distributed through out the country, where sorghum is grown.

Nature of damage : The damage is caused by larvae which feed initially on lemina by scraping the chlorophyll and lateron on the milky as well as hard grains of sorghum. The caterpillars web together adjacent grams in a silken thread and feed on the grains inside the earheads. Heavy infested grains found to have more webs and excreta of the pest. The damage is caused generally from September to November.

Life history : The four developmental stages, *viz.*, egg, larva, pupa and adult are found in its life cycle which have been described below.

Egg : The female lays eggs on spikelets and on grains of the earheads. Freshly laid eggs are creamy unite in colour, round or conical, flattened in shape. The individual egg measures 0.45 mm in length and 0.32 mm in breadth. They are laid singly. The incubation period varies from 3-4 days.

Larva : The newly hatched caterpillar is dirty white in colour with brown head. Body is translucent without strips consisting of 12 segments. They feed on the developing grains and grow in size. The full grown larva is active measuring about 12 mm in length and 2 mm in breadth with dark brown colour. The larvae form web come out of the excreta and the silken thread on the cob and remain inside. The larval period lasts for 9-10 days.

Pupa : Pupation takes place in the silken webs. The pupa is short and brown in colour, covered with silken membranous webbing threads which are secreted by the caterpillar. The pupa measures about 7.5 mm in length and 1.8 mm in breadth. The pupal period lasts for 7.8 days.

Adult : The adult measures 6.5 mm in length with using expanse of 7 mm. The thorax is dark grey in colour. The hind wings are having fringe like hairs on anterior margin and are transparent. The fore wings are smaller than hind wings and dark-grey in colour. The longevity of male and female moth in 3 and 5-6 days, respectively.

The newly emerged adults mate in the morning hours and start egg laying after 20-24 hours of mating. The total life cycle is completed in 20-22 days and there are 4-5 generations in a year.

Control : Spraying the crop with any one of the following insecticides :

(i) Endosulfan — 0.07%

(ii) Monocrotophos — 0.05%

(iii) Phosphamidon — 0.03%

4

Crop Destroyers

Kharif Grasshopper

Local name : Phangi, Kata, Tota, Pharka, Ankhfura and Tidda.

Scientific name

(i) *Hieroglyphus nigrorepletus Bolivier*

(ii) *Hieroglyphus banian Fabr.*

Order : Orthoptera

Family : Acrididae

Host plants : Paddy, Jowar, Bajra, Maize, Sugarcane, Smaller millets, Sunhemp, Arhar and Grasses.

Distribution : Besides India, it is found in Mayanmar, Sri Lanka, Malaya and Hawaii islands. In India it is distributed in all paddy growing areas.

Nature of damage : Pest appears in the field on the, onset of monsoon. Initially it feeds on grasses and as soon as Kharif crops are grown, it starts feeding on them. The damage is caused by both adults and nymphs having chewing and cutting type of mouth parts. In comparison to adults, nymphs cause more damage because nymphal period is longer varying from July to September and secondly they eat more to fulfill the various activities (growth etc.) of the life while adults feed only for their survival.

They feed on leaves and shoots and in case of severe infestation they completely defoliate the plants leaving only the mid-ribs. Due to which the growth of plants badly affected and remains stunted resulting in poor yield. At the time of milky stage in rice, they also feed on ears thus causing heavy loss. The severe damage is usually done during August and September. Though it is a sporadic but serious pest in most of the eastern Uttar Pradesh. Few years back it appeared in epidemic form in Rajasthan and completely destroyed the crops. It has been estimated that an individual grasshopper during its life time eats about 42 grams of green maize leaves.

Life cycle : Three developmental stages, *viz.*, egg- nymph and adult, are found in its life cycle.

Egg : The eggs are deposited by the female in masses of 30-40 at a depth of 5 to 7.5 cm in the soil along the side of field bunds from October to December. When the female is ready to lay eggs it trails its abdomen on the ground for searching of suitable place. After finding a suitable place the female drills the surface of the soil with its ovipositor and forms a hole for egg laying. When hole is ready it secrets a fluid from gonopore and starts egg laying one by one in it. The abdomen is withdrawn as the hole gets filled with eggs. This hole is called as egg chamber.

The female again secretes a fluid on the eggs and then filled up the egg chamber with the loose soil by its hind legs. The foamy material exuded by the female serves for sticking the eggs together and for cementing the soil particles which constitute a pod-like structure with eggs within it. These egg masses are known as egg pod which remain in the soil till the onset of rains in June-July of the following year. Generally single female lays 15 to 20 egg masses in her life span. An individual egg is light yellow in colour and resembles to paddy grain in shape. At the time of hatching colour changes in to brown. The egg period is of about 9 months.

Nymph : The newly hatched tiny numphs are light yellow in colour measuring 6 mm long and devoid of wings. They start feeding on grasses and can be seen jumping from one place to another. The nymphal period is of 70 to 80 days and during this period they cast their skins 5 or 6 times at an interval of 10 to 15 days. After second moult the colour changes to light green and wing pads are visible.

Adult : The adult hoppers are yellowish-green in colour. *H. nigrorepletus* is larger than *H. baninan;* the male and female of the former are 37 mm and 75 mm long while of the later 25 mm and 37 mm long. In case of *H. banian* there are only 2-3 black markings but in *H. nigrorepletus* there is a reticulum of black lines and possesses U shaped black spot on the lateral sides of pronotum.

The adults are seen in the last week of September or early October. They copulate and may be seen for hours in same position. The males die after copulation while females after egg laying. Only one generation is found in a year. After becoming adults the insects take 7 to 20 days to attain sexual maturity and after the first mating they require a period of about 7 days to lay the first batch of eggs. The life of female adult lasts for about two months.

Control

1. Plough the affected fields and the bunds to a depth of 10-15 cm before the monsoon begins, it will destroy the egg masses.

2. The hoppers and adults can be collected and destroyed.
3. At the time of on set of monsoon they may be attracted by light-traps and destroyed.
4. The dusting of following insecticides has been found effective in minimising the population of pest :
   (a) Methyl parathion 2% (3) 30 kg/hect.
   (b) Fenitrothion 5% @ 20 kg/hect.
   (c) Malathion 5% @ 30 kg/hect.
5. The spraying of following insecticides has also been found quite effective :
   (a) D D V P 76 EC-200 ml/hect.
   (b) Malathion 50 EC-2.5 lit/hect.
6. Poison bait with following ingredients may also be prepared and spread in the field to control the hoppers:
   (a) Bran-10 kg.
   (b) Sodium arsenate- 1/2 kg.
   (c) Mollases-2.5 lit.
   (d) Water-10 lit.

Natural enemies : The following are the egg parasites of this pest :

(a) *Cacallus sp.*

(b) *Barycomus sp.*

(c) *Scelio sp.*

Some birds like crow and kite eat the nymphs and adults while others like wood-pecker and blue-jay eat the eggs of the hoppers.

Gundhi Bug

Local name : Rainya, Chusia, Gundhi

Scientific name : *Leptocorisa varicornis* Fabr.

L. costalis

L. acuta

Order : Hemiptera

Sub-order : Heteroptera

Family : Coreidae

Host plants : Paddy, jowar, bajra, maize, smaller millets and many grasses.

Distribution : It is found in most of the rice-growing areas of India and its distribution extends over the whole of South-Asia from China and Japan in the north to Sri Lanka in the south. It is also found in the Philippine islands and Australia. In India its attack is specially severe in Uttar Pradesh and Bihar.

Nature of damage : It is the most serious bug pest of the paddy crop. Both nymphs and adults suck the juice initially from the leaves and tender shoots and later from the ears. Although it feeds on a variety of plants but creates serious problem when grain-formation is in progress in the paddy. It sucks the milky juice, leaving the white chaffy husk. A hole is left on the grain where bug punctures with its proboscis and around which a brownish spot develops due to fungus attack. Its attack can easily be recognised by unpleasant odour which comes from the field where it is present and the due to which name 'gundhi' has been given. The insect is very common i.e. paddy field from middle if August to middle November and causes 20 to 30% damage of crop. In case of severe infestation, practically all the grains of all the earheads become chaffy. During 1958 it appeared in epidemic from in Assam. West Bengal, Bihar, Madhya Pradesh and U.P. and control operations where under taken in 4,84,168 acres of paddy fields in U.P.

Life cycle : Three developmental stages namely egg, nymph and adult are found in its life cycle which are described as under :

Egg : The eggs are laid by the female bug in the month of July and August. They are deposited in three stripes on the lower side of the leaves. Each stripe contains 10 to 20 eggs and about 40 eggs are laid at the place. The eggs are blackish-brown small bead like in structure. Depending on the prevailing temperature, the eggs hatch in about 4-7 days. At the time of hatching their colour changes into black.

Nymph : The slender greenish nymphs hatch out from the eggs, start immediately sucking the plant juice. At the time of hatching it is about 1.8 m long which grows up to 2.0 mm within 6 hours. There are five moults after which the nymph becomes adult with wings. The wings-pad may seen in the IIIrd nymphal instar. The nymphal period is of 14 to 20 days. The full grown nymph measures 15 mm long.

Adult : The adult bug is about 15-16 mm long slender and greenish-yellow in colour. Head is small and provided with 4 segmented rostrum. The antennae are larger than the body and red in colour. Meta-thorax provided with glands emitting unpleasant odour. The fore wings are hemelytrat.

The life cycle is completed in 21 to 31 days and several overlapping generations are found but on paddy 4 to 5 generations are completed. The bugs copulate after 12-14 days of emergence and female survive up to 55 days while male only for 33 days. The pest passes its winter in adult stage and migrate to grasses and other plants to continue it's life cycle.

Control

1. All weeds and grasses growing around the paddy field should be destroyed.
2. The nymphs and adults may be attracted in light-traps.
3. Resistant varieties should be preferred in infested areas,

e.g., Sathia in U.P., Sona in Bihar and Mundga kuttin in Chennai.

4. The dusting with any one of the following insecticides has been found quite effective :
   (a) Methyl pasathion 2% dust @ 25 kg/hect.
   (b) Malathion 5% dust @ 30 kg/hect.
   (c) Thiodan 4% dust @ 20 kg/hect.
   (d) Carbaryl 10% dust @ 20 kg/hect.

Natural enemies : The important natural enemies of the pest are given as follows :

(a) Mantis-feeds on nymph and adults.
(b) The following coccinellids also feed on nymphs and adults :
   (i) Coccinella undecipunctata
   (ii) Adalia sp.
   (iii) Ommatius sp.
(c) Pentatomid and reduvid bugs also prey on them

Fruit Fly

Common name : Son Makhi, Fal mashi, Fal makhi etc.

Scientific name : Dacus (Strutmeta) cucurbitae Coq.

Order : Diptera

Family : Trypetidae

Host plants : Cucurbitaceous plants such as gourd, pumpkin, cucumber, Kerala, taroi, melon, khera, tinda and some times tomato, papaya, brinjal and guava.

Distribution : The fruit flies are widely distributed through out the world. It was first seen in Hawaii Island during 1898 and by 1913 spread all over the world.

Nature of damage : The damage is caused by the maggots of the fly and considered to be a group of very harmful insects which cause serious damage to fruits and vegetables. Generally, fruits are attacked in the early immature soft stage with the result that; they do not develop, drop down and rot. Those which do not drop and are picked up, found unfit for human consumption as they have a number of abominable white maggots in the pulp. The humid-hot weather is most suitable for its attack and generally 50 to 60 percent loss is caused to vegetable crops.

Life cycle : There are four developmental stages in its life cycle which are described as follows:

Egg : The female fly lays eggs from February to November. The pre-oviposition period is 14 days in August and 9 to 30 days during other months depending upon the environmental conditions. The fly lays small, shiny whitish cigar shaped eggs in small clusters just under the skin of the fruits with the help of her ovipositor and measuring 0.6 mm long. Each cluster may have 4 to 10 eggs and they are laid in many times at an interval of 2 to 5 days. Female lays about 300 eggs in her life span. The hatching periods of eggs varied from 2 to 9 days.

Larvae (Maggots) : The maggots on hatching feed on the fruit pulp by burrowing into it. They are white in colour and measuring 1 mm long. Their body is divided into 11 segments and one end of the body is wider than the other. They are devoid of legs. The maggots spend their whole larval period in side the fruit and by this time, the attacked fruits are generally rot and the larvae leave the rotten fruit and enter the soil for pupation. The larval period takes 7 to 8 days but may be 3 days in summer and 20 to 21 days in winter. Full grown maggot is 19 mm long and 2 mm wide.

Pupa : Pupation takes place in the soil few cm below the surface. The pupa is flat, oval pointed towards anterior and flat to posterior end. It is 11 segmented and 5.5 mm long with light yellow in colour. The pupal period generally lasts 6-9 days which may extend 20-21 days in November and 28-30 days during December.

Adult : The fly is reddish-brown in colour with black and white spots. The wings have brown bands and spots at the apex. It measures about 7 mm in length and 3 mm in breadth with a wing expansion of about 14 mm. The head is like a top provided with aristate antennae. There are 3 bright yellow stripes on the meta-thorax.

The life cycle is completed in 15-20 days during summer and rainy season which extends to 60 days in winter. There are many overlapping generations in a year. The adults are generally long lived; male survives about 56 days while female for about 66 days. They remain in hibernation during November to January.

Control

1. The ploughing of fields two to three times reduces the pest population by exposing the pupae to enemies.
2. Collection and destruction of infested fruits along with the larval stage of the fruit-fly.
3. Baging of fruits protects the fly from egg-laying.
4. Dusting the plants with tobacco dust or ash mixed with kerosine oil to repel the flies.
5. Male flies may be attracted in citrenela oil and destroyed.
6. The adult flies may also be controlled with the spray of following solution :
   (a) Tartar acid-1 part
   (b) Water-320 parts
   (c) Mollases- 24 parts

7. Use bait-traps prepared with the following. This can be put in small tin containers which are hung from sticks at various places in the field :

    (a) Protein hydrolysate-100 g.

    (b) Water-10-20 lit.

    (c) Malathion 25% W.P.-200 g.

8. Some chemosterilants, *viz.*, tepa, metepa and apholate have been found effective in sterilizing the fly which will prevent the multiplication of pest.

9. The following insecticides may be sprayed regularly during the active period of the fly.

    (a) Malathion 50 EC/W.P.-2.0 lit or 2.0 kg/hect.

    (b) Formothion 25 EC-0.650 lit/hect.

    (c) Carbaryl 50 W.P.-3.0 kg/hect.

    (d) DDVP 76 EC-0.250 lit/hect.

Natural enemies : The following are the pupal parasite of the pest :

(a) *Opius fletcheri* Silv.

(b) *O. incisus* Silv.

(c) *O. compensans* Silv.

Red Pumpkin Beetle

Local name : Torai ka kira, Lal kira

Scientific name : Three species of the insects are common:

(a) *Raphidopalpa foveicollis Lucas* — *Red coloured*

(b) *R. intermedia Jaobx.* — *Black round*

(c) *R. cincta Fabr.* — *Brown coloured*

Order : Coleoptera

Family : Chrysomelidae

Host plants : All the cucurbitaceous plants such as melon, cucumber, gourd, torai, tinda, kheera, and kakari, etc.

Distribution : Besides India; it is distributed Ceylon, Pakistan, Mayanmar, Sudan, Italy and Australia. In India, it is most common in U.P., Punjab. M.P. Maharashtra, Bihar and Rajasthan.

Nature of damage : The damage is caused by both grubs and adults of the insect. The adult beetles feed on the leaves, flower buds and flowers, there by causing severe injury specially to young plants. The damage done to young seedlings is often so severe that the crop has to be resown because all the leaves of young plants are eaten up. In older vines the beetles eat and make holes into the leaves. The grub bores in to the root and stem of the plant and destroy the whole plant when present in large number. The attacked plants wither and die. They also attack the fruits near the portion resting on the ground and become unmarketable.

Life cycle : Four developmental stages namely egg, larva (grub), pupa and adult are found as described below :

Egg : Hibernating beetles come out in the beginning of March and copulate after 7 to 8 days of emergence. Pre-oviposition period is of 2-7 days. The females deposit orange-coloured eggs singly or in small clusters on moist soil near the base of the seedlings. A single female is able to lay 60-300 eggs in her life span. The incubation period of the eggs varies from 5-15 days depending on environmental temperature and humidity.

Grub : The grubs soon after hatching are quite active and begin to feed on the roots and stems which readily bore in to. They are creamy-yellow in colour with brown head and prothorax. The last segment is dark coloured and is provided with a medium cylindrical process which is used by the grub in walking. At the end of every instar, the larvae enter the soil to moult and come out there after to feed again. There are four larval instars before it becomes full grown in 13-25 days. The full grown grub is about 15 mm long.

Pupa : At the end of the last instar when the larva is full grown, it tunnels in to the soil to depths varying from 1 cm to 2.5 cm and there it prepares an oval chamber for pupation. The pupa is reddish-brown in colour and measuring 7.5 x 3.5 mm. The pupal period varies from 7 to 14 days.

Adult : The adult beetle measures about 7 mm in length and 4 mm in breadth. The upper surface of the body is bright orange-red while lower surface of the abdomen is entirely black and covered with short whitish soft hairs. The posterior portion of the head is covered with pro-thorax.

The adult beetle soon after emergence begins to feed and starts egg laying after a week. The life cycle is completed in 32-61 days and there are 5 to 8 generations in a year. During winter the beetles remain in hibernation hiding in various niches from the severity of the cold. As soon as it begins to warm up, the over wintering population becomes active and comes out from its hiding places.

Control

1. Plough the fields after harvest to destroy the pupae and larvae therein.
2. Early sowing of cucurbitaceous plants, *i.e.*, in November protects the crop from the beetles appearing after hibernation as the plants are well established in the mean time.
3. Dusting the crop with kerosinized ash, will repel the beetles.
4. Interculturing around the roots of vines in the month of March-April minimises the attack of grubs.

5. The vines may be dusted with any one of the following insecticides :
   (a) Malathion 5% dust @ 15-20 kg/hect.
   (b) Carbaryl 5% dust @ 20 kg/hect.
   (c) Endosulfan 4% dust @ 15-20 kg/hect.
6. Sprayings with following insecticides have also been found effective :
   (a) Malathion 50 EC/Wp- 1.25 lit or 1.25 kg/hect.
   (b) Formothion 25 EC-0.650 lit/hect.
   (c) D D V P 76 EC-0.250 lit/hect.

Natural enemies : Some birds like crow, myna and blue jay devour the larvae and adults of the pest. Preying-mantis has also been observed preying on adults.

Epilachna Beetle

Scientific name : *Epilachna vigintioctopunctata* (Fabr.)

E. dodicastignia (Mulsant)

Order : Coleoptera

Family : Coccinellidae

Host plants : Potato, brinjal, tomato, bitter gourd and other cucurbitaceous and solanaceous plants.

Distribution : Through out India.

Nature of damage : The damage is caused by both adults and grubs which feed on the upper surface of the leaves. They skeletonize the leaves which present a lace-like appearance as the green matter in between the veins is eaten away leaving the skeleton of anastomosing veins. They eat up regular areas of the leaf tissue, leaving parallel bands of uneaten tissue in between. The leaves turn brown, dry up and fall off, completely detoliating the plants.

Life history : The four developmental stages namely egg, larva, pupa and adult is found in its life cycle which have been described below :

Egg : The eggs are generally laid on the under surface of leaves in a prominent yellow batches of 5-40 eggs. They are cigar shaped and glued to the leaf surface in a vertical position. A single female can lay upto 400 eggs in her life span. The hatching period of eggs varies from 3-4 days which may extend upto 9 days in winter.

Larva : The newly hatched grub is stout and yellow in colour, bears spines all over the body. They feed on the lower opidermis of the leaves. The full grown grub measures 8 mm in length and 4 mm in breadth, the body being brood in front and narrow behind and covered with spiny structures all over. The larval period varies from 12-18 days during which there are three larval instars.

Pupa : The pupation takes place on the leaf surface or around the base of the stem. At that time the full grown larva attaches the last segment of its abdomen to the leaf surface by means of a sticky secretion and the pupa is formed within the last larval skin which splits on the dorsal side. The pupa is yellowish-orange in colour with brownish white markings on the upper surface. The pupal period varies from 3-6 days.

Adult : The adult is yellowish-brown in colour convex, oval shaped and bears small, round black spots on the elytra. The *E. vigintioc-topunctata* bears 14 such spots on each elytra while *E. dodicastigma* possesses 6 spots on each elytra. Adult longevity varies from less than a month to more than two months. The beetle passes winter in hibernating stage in the heaps of dry plants or cracks and crevices in the soil.

The total life cycle is completed in 19 to 30 days and there are 7-8 generations in a year, while in hills only one generation is completed. The pest remains active from March to October and during hot and dry months, the number decline very much, but the population builds up again in August.

Control

1. Collect and destroy the egg masses and other stages of pest during early stage of infestation.
2. Dust the crop with any one of the following insecticides per hectare :

   (i) Malathion 5% dust - 25 kg/hect.

   (ii) Carbaryl 5% dust - 30 kg/hect.

   (iii) Endosulfan 2% dust - 25 kg/hect.
3. Spray the crop with any one of the following insecticides:

   (i) Endosulfan - 0.05%

   (ii) Methyl parathion - 0.02%

   (iii) Malathion - 0.05%

Brinjal Shoot and Fruit Borer

Scientific name : *Leucinodes orbonalis* Guen.

Order : Lepidoptera

Family : Pyralidae

Food plants : Brinjal and other solanaceous plants.

Distribution : The pest is widely distributed in Malaysia, Sri Lanka, India, Pakistan, Germany and East Africa.

Nature of damage : The damage is caused by the caterpillar which attacks the top shoots of young plants as well as fruits. The larvae bore into the shoot and eat the internal tissues causing the attacked portion to droop and wither away. When the terminal shoots are attacked the growing points are killed. In young fruits, the newly hatched larvae bore into the fruit and the entry hole is too small to be easily noticeable. The large holes usually seen on the fruits are the exit holes of the caterpillars. The infested fruits become unfit for human consumption. Damage to the fruits, particulary in the autmn is very severe and the whole crop may be destroyed by the borers. A single caterpiller may destroy 4-6 fruits.

Life history : The four developmental stage *viz.*, egg, larva, pupa and adult ore found in its life cycle which have been described below :

Egg : The female moth lays eggs singly or in batches of 2-4 on the underside of leaves, on green stems, flower buds or the caly of fruits. The eggs are creamy-white in colour and elongate. Each female generally lays more than 150 eggs. The hatching period of eggs varies from 3-6 days.

Larva : The young caterpillars bore into tender shoots near the growing pointer, into the flower buds or into the fruits. The young caterpillar is creamy white in colour which becomes pale white with violet mashings arranged in linear lines all over the body when full grown. The full grown caterpillar measures about 18-23 mm in length. The larvae grow through 5 stages and larval period losts for 14-20 days.

Pupa : The full grown caterpillars comes out of the fruits or shoots and pupate amongst fallen leaves in a boat shaped silken cocoon. The pupal period lasts for 7 to 11 days after which the moths emerge.

Adult : The moths are medium sized insects with white colour but there are pale brown or black spots on the dorsum of thorax and abdomen. Its wings are white with a pinkish or bluish tinge and are ringed with small hair along the apical and anal margins. The moth measures about 20-22 mm across the wings. The fore wings are ornamented with a number of black, pale and light brown spots.

The total life cycle is completed in 25-39 days and there are 5 overlapping generations in a year. The caterpillars hibernate in the winter and pupate early in the spring. The moths appear in March-April to continue the next generation.

Control

1. Collect and destroy the drooping shoots and affected fruits which contain caterpillars inside.
2. Spray the crop with any one of the following insecticides:
   (i) Endosulfan - 0.07%
   (ii) Carbaryl - 0.15%
   (iii) Malathion - 0.05%
   (iv) Fenvalerate - 0.003%

Tobacco Caterpillar

Scientific name : *Spodoptera litura* Fabr.

Order : Lepidoptera

Family : Noctuidae

Host plants : Tobacco, tomato, castor, mustard, cabbage, cauliflower, knolkhol, potato, gram, pea and chillies etc.

Distribution : It is found throughout the tropical and sub-tropical parts of the world. The pest is widely distributed in India.

Nature of damage : The damage is caused by the caterpillar which feed on the leaves and fresh growth of the plants. They are voracious feeder of leaves and several of them when present on a plant cause defoliation. The pest in active in the night and feeds by biting large holes in the leaves. Some times whole crop is destroyed due to their attack particularly in tobacco nurseries.

Life history : The four developmental stages, *viz.*, egg, larva, pupa and adult are found in its life cycle which are described below.

Egg : The moths are active at night when they mate and female lays eggs in cluster on the under surface of the leaves. The eggs are covered with buff-coloured hairs. Each cluster may have 100-250 eggs. The hatching period of eggs varies from 4-5 days.

Larva : The newly hatched larvae feed gregariously for first few days and then disperse to feed individually. The full grown caterpillar is about 35-40 mm long and is velvety black with yellowish-green dorsal stripes and lateral white bands. They pass through 6 stages. The larval period lasts for 15-30 days.

Pupa : When the larva is full grown, it comes down off the plant and pupates in the soil. The pupa is reddish brown in colour and measures 20-22 mm in length. The pupal period lasts for 7-15 days.

Adult : The moth is a medium size, dark brown in colour with irregular whitish lines on the lore wings. The hind wings are white. They are about 22 mm long and measure 40 mm across the spread wings. After emergence they live for 7-10 days.

The life cycle is completed in 27 to 52 days and there are about 8 generations in a year. The pest breeds throughout the year, although its development is considerably reduced during the winter.

Control

1. During early stage, when they feed gregariously may be collected and destroyed.
2. Spraying the crop with anyone of the following insecticides :

(i) Malathion – 0.05%

(ii) Carbaryl – 0.15%

(iii) DDVP – 0.03%

(iv) Endosulfan – 0.07%

Cabbage Butterfly

Scientific name : *Pieris brassicae* Linn.

Order : Lepidoptera

Family : Pieridae

Host plants : Cabbage, cauliflower, knolkhol, turnip, radish and mustard etc.

Distribution : The cabbage butterfly is world-wide and is found wherever cruciferous vegetables are grown.

Nature of damage : Damage is caused by the caterpillar which feeds on the leaves of the plant. The first instar caterpillars just scrap the leaf surface, whereas the subsequent instars eat up leaves from the margin inwards leaving intact the main viens. When the caterpillars are abundant they leave only the bare stalks of the leaves on the plants.

Life history : The four developmental stages, *viz.*, egg, larva, pupa and adult are found in its life cycle which are described below :

Egg : The female butterfly is vary active in the field and lays a large number of eggs on the leaves of food plant. On an average, 164 yellowish conical eggs in clusters 50-90 eggs are laid on the lower or upper surface of the leaves. The hatching period of eggs varies from 3-7 days in March-May and 11-17 days in November to February.

Larva : The small green caterpillars with black heads come out from the eggs and start feeding on the leaves. The caterpillars feed gregariously during the early instars and disperse later on. The larvae are pale-yellow, when young and become greenish yellow when full grown. The head is black and the dossum is marked with black spots. The full grown caterpillar measures about 40-50 mm in length. They moult four times and become full grown in 15-22 days during March-April and in 30-40 days during November to February.

Pupa : The full grown caterpillar pupates either on a leaf or on a tree or at any appropriate place with a girdle of silk round the middle to hold the chrysold in position. The pupal stage lasts 8 to 15 days in March-April and 20-28 days in November to February.

Adult : The adults are pale white butterflies and have a smoky shade on the dorsal side of the body. The wings are pale white, with a black patch on the apical angle of each fore wing and a black spot on the costal margin of each hind wing. The male is small than female. The female below larva measure 6.5 cm across the spread wings. They live for 2-12 days depending upon the season.

The pest appears on cruciferous vegetables in the beginning of October and remains active upto the end of April. It is not found in plains from May-September during these months it breads in the hills. The life cycle is completed in 27-87 days and there are 4 generations in a season (October to April).

Control

1. Collection and destruction of egg masses and larvae during early stage.
2. Spray the crop with any one of the following insecticide:
   - (i) Malathion - 0.05%
   - (ii) Endosulfan - 0.07%
   - (iii) Carbaryl - 0.15%

Diamond Back Moth

Scientific name : *Plutella xylostella* Linn.

Order : Lepidoptera

Family : Plutellidae

Host plants : Cauliflower, cabbage and rape-seed etc.

Distribution : It is world-wide in distribution and found throughout Indian union.

Nature of damage : The damage is caused by the caterpillars which is in the earlier stages, feed in mines on the lower side of the cabbage leaves and in the later stages on every position of the leaves. Central leaves of cabbage and cauliflower may be riddled and the vegetables rendered unfit for human consumption. They particularly feed in the heart or head of cabbage and cauliflower. It is most serious when appears in the early stages of the crops, *i.e.*, during August-September.

Life cycle : The four developmental stages, *viz.*, egg, larva, pupa and adult are found which are described below :

Egg : The eggs are laid by the female moth on the under side of the leaves. They are usually laid singly or in batches of 2-40, which are yellowish in colour. The single female can lay about 356 eggs in her life span. The hatching period of eggs varies from 3-9 days.

Larva : The newly hatched larvae bore into the tissues from the underside of the leaves and feed in the tunnels. In the third instar, the caterpillars feed outside the tunnels and in fourth instar feed from the underside of the leaves, leaving intact a parchment-like transparent cuticular layer on the dorsal surface. The full grown larva measures about 8 mm in length and are pale-yellowish-green, with fine black hair scattered all over the body. The larval period lasts for 8-16 days.

Pupa : The full grown larva constructs a barrel-shaped silken cocoon, which is open at both ends and is attached to the leaf surface. The larva pupates inside the cocoon and comes out in 4-5 days as an adult.

Adult : The adult measures about 8-12 mm in length and is brown or grey in colour. They have conspicuous white spots on the fore wings, and appears like diamond patterns when the wings lie flat over the body. The moths may live for as long as 20 days.

The pest is active throughout the year. The life cycle is completed in 17-23 days and there are several generations in a year.

Control

The pest may be controlled by the sprayings of anyone of the following insecticides at weekly intervals :

(i) Malathion - 0.05%

(ii) Endosulfan - 0.07%

(iii) Fenitrothion - 0.05%

(iv) Carbaryl - 0.15%

Mustard Aphid

Local name : Chenpa, Mahun, Mahon, Tela, Mawa, Moyala, Lahi.

Scientific name : *Lipaphis erysimi* Kalt. *(Siphocoryne indobrassicae)*

Order : Hemiptera

Sub order : Homoptera

Family : Aphididae

Host plants : Mustard, cauliflower, cabbage, knolkhol, radish and turnip etc.

Distribution : It India it is found in all the states specially in U.P., M.P., Punjab, Maharashtra, Bihar, Gujarat, Mysore and Rajasthan.

Nature of damage : In appears in the last week of November or early December in U.P. and reaches to its peak in the month of February. Both nymphs and adults suck the sap from the tender portion of the plants. Hundreds and thousands of these may be seen on leaves or tender shoot. In mild case, the shoot wilts and in severe cases it completely gets killed. The honey-dew produced by such a large number of individuals covers practically the whole surface of the tender leaves and shoot and a kind of black mould develops which interferes the photosynthetic activities of plant. Honey-dew secreted by aphids is liked by a number of other insects specially by ants who for the sake of ensuring the supply of this wet liquid, protect the aphids and also provide them transport; this type of association is known as commensalism. The cloudy and humid atmosphere is most suitable for its multiplication and attack.

Life cycle : A different type of life-cycle has been found in plain and hill regions. In plains, the insect does not lay eggs hence only nymphs and adults are found.

In plains : They appear in November-December on mustard plants and among them females are more in number than males. The females are viviparous and they do not lay eggs but produce many nymphs. The nymphs are produced by following ways :

(a) After mating the males.

(b) Without mating the males but parthenogenetically.

The nymphs become adult within 3 to 7 days and start producing apterous offsprings. In this way they increase great in number and spread over a big area in a very short time. There are several generations during the cold season.

In hills : In hills the life cycle of aphids is just like in European country. Aphids are remarkable on account of their peculiar mode of development and the polymorphism exhibited in different generations of the same species. The following types of individuals are present in the life cycle of migratory aphids :

Fundatrices : These are apterous, viviparous parthenogenetic females which emerge in spring from the over-winter and eggs. Only one generation is found of this group.

Fundatrigeniae : These are also apterous, parthenogenetic, viviparous females and are the progeny of fundatrices. They live on the primary host and complete 3 generations.

Migratory : These usually develop in the second, third or later generations of fundatrigeniae and consist of winged parthenogenetic, viviparous females. They develop on the primary host and subsequently fly to the secondary host.

Alienicolae : Parthenogenetic, viviparous females developing for the most part on the secondary host. They often markedly from the fundatrices and migrants. They are similar to fundatrigeniae but differ from them by living on secondary host while former on primary host. Many generations may be produced comprising both apterous and winged forms.

Sexuparae : Parthenogenetic, viviparous females which develop on the secondary host and migrate to the primary host at the end of summer.

Sexuales : These are the progeny of sexuparae produced on the primary host. They consist of sexually reproducing males and females. The females lay eggs on the primary host after copulating the males which passes winter as such and hatch out in following spring and fundatrices females are produced.

Adult : The aphids are small generally 2 mm in size, green sucking insects. They are provided with a pair of small tubular structure projecting out from dorsal surface of the posterior region of the body known as cornicles or siphons or honey tubes. Two pairs of transparent wings are found in which costa and sub-costa veins are absent.

Control

1. Early sowing or early maturing varieties should be grown to escape the damage.
2. Resistant varieties *viz.,* AGH-4 and T. 101 should be preferred to grow.
3. The crop may be dusted with any one of the following insecticides :
   (a) Malathion 5% dust @ 25 kg/hect.
   (b) Carbaryl 10% dust @ 25 kg/hect.
   (c) Fenetrothion 5% dust @ 20 kg/hect.
4. The spraying of any one of the following insecticides has been found effective to control the aphids :
   (a) Phosphomidon 100 EC @ 0.250 lit/hect.
   (b) Dimethoate 30 EC @ 1.00 lit/hect.
   (c) Endosulfan 35 EC @ 1.25 lit/hect.
   (d) Formothion 25 EC @ 0.650 lit/hect.
   (e) Methyl demeton 25 EC @ 1.00 lit/hect.
   (f) Malathion 50 EC @ 0.600 lit/hect.
   (g) Chlorpyrifos 20 EC @ 0.600 lit/hect.

Natural enemies : The following insects feed on aphids :

(a) *Coccinella septempunctata (Lady bird beetle)*

(b) *Chrysopa sp. (Aphid lion)*

(c) *Syrphid sp. (Maggots of syrphid)*

Mustard Saw-fly

Common name : Sarson ki illi

Scientific name : *Athalia proximo* Klug.

Order : Hymenoptera

Family : Tenthredinidae

Host plants : Mustard, radish, turnip, cabbage, cauliflower and other cruciferous plants.

Distribution : The pest is distributed practically throughout India. Besides India, it is widely found in Indonesia, Formosa and Mayanmar.

Nature of damage : In most parts of the country it is pest of cold weather and its activity confined generally to the period from October to March although in certain western regions it is reported to be prevalent round about August. The larval stage damages crop by voractiously feeding on the young crop. Its feeding activities are generally confined to mornings and evenings. As a result of feeding the leaves become full of holes and the plants either dry up or remain stunted. In case of severe infestation all the leaves are eaten and the plants die.

Life cycle : Four developmental stages namely egg, larva (grub), pupa and adult are found in its life cycle.

Egg : The female lays eggs singly in to the tissues of the food plant. The egg laying organ (ovipositor) is highly specialized for silting open the margin of the leaves within which eggs are laid. Initially they are milky in colour which turn in to dark brown or black at the time of hatching. An egg is 2 mm long and many eggs may be laid in to one leaf. Each female is capable of laying usually 30-61 eggs which in some cases may reach up to 150 eggs. Hatching period of eggs varies from 4 to 27 days depending upon the environmental conditions.

Larva (grub) : The young grub is greenish grey in colour and its body surface is hairless measuring 2 mm long. It begins to feed on the margin of the leaf, and as it grow, its colour getting darker and darker. Full grown grub is 15 to 20 mm long having 3 pairs thoracic and 8 pairs abdominal legs and provided with 5 stripes on the back. The larval period lasts for 12 to 18 days.

Pupa : For pupation grub goes in to soil wherein it prepares cocoon made of silk buttressed by soil particles. The cocoon is 7-11 mm long and 4 to 6 mm wide. The pupal period is of about 10-12 days. During severe cold it hibernates in pupal stage.

Adult : The adult sawfly has a short thick-set body with a mixture of yellow brown markings on it and a dark reddish brown colour on the wings. It has two pairs of wings and good flier, its activities are diurnal. The fly is also capable to lay eggs parthenogenetically, but only males emerge out from these eggs.

Control

1. Hand picking of grubs early in the morning is found useful to check the population of the pest.
2. Dusting the crop with any one of the following insecticides has been found effective :
   (a) Malathion 5% dust @ 25 kg/hect.
   (b) Thiodan 4% dust @ 30 kg/hect.
   (c) Carbaryl 5% dust @ 25 kg/hect.
3. The crop may also be sprayed by any one of the following insecticide to destroy the pest :
   (a) Dimecron 85% @ 0.250 lit/hect.
   (b) Thiodan 35 EC @ 1.0 lit/hect.
   (c) Nuvan 100 @ 0.25 lit/hect.

4. 0.1% spray of brestnol 45 WP protect the crop from grubs. It acts as antifeedant due to which the larvae are unable to locate its host and finally die.

Natural Enemies

Larval parasite : *Exacrodus populence.*

Painted Bug

Local name : Sundar, Jhanga, Dagila Keet

Scientific name : *Bagrada cruciferanum* Krik.

Order : Hemiptera

Sub-order : Heteroptera

Family : Pentatomidae

Host plants : Mustard, cabbage, cauliflower, knolkhol, radish, turnip and other cruciferous plants.

Distribution : It is found throughout India and besides India distributed in Mayanmar, Sri Lanka, Iran and East Africa.

Nature of damage : Both nymphs and adults cause the damage to the crop by sucking the sap from the leaves, stems and tender parts of the plants. The attack of pest is noted after few days of sowing the crop, if they are in great number may destroy the whole crop and necessitate resowing. The attacked plants look sickly and dry up or may get stunted in growth. The black fungus is also attracted at the feeding point due to which brown or black spots are seen on the leaves. In the later stage of the crop they suck the sap from the flowers and pods resulting in the poor formation of pods and yield. After harvesting pest may be seen in greater number on threshing floor.

Life cycle : Three stages, *viz.*, egg, nymph and adult are found in its life cycle.

Egg : The eggs are laid by female but usually singly on the leaves, leaf stalks and stems of the host plants. Some times female also lays eggs in soil near the roots in cluster. The number of eggs laid by each female varies from 90-250. Freshly laid eggs are pale yellow in colour which turn in to pinkish at the time of hatching. The individual egg measures 1 mm x 1/2 mm. Hatching period of eggs varies from 3 to 7 days.

Nymph : The newly hatched nymph is small, bright orange in colour and with dark red eyes. They are 1.3 mm in length and 1 mm in breadth. They start feeding on the plants and develop rapidly. Nymphs moult 5 times and attain the winged adult stage. Full grown nymph measures 4.5 mm x 3.2 mm. The nymphal period varies from 20-27 days.

Adult : The adult bug is flat, bright black in colour with reddish yellow spots and dots and measures 5 mm long. The head is small, triangular and provided with black eyes, two ocelli and setaceous antennae. The scutellum is very large.

The life cycle is completed in 30-41 days and 6 to 7 generations are found in a year. It is most abundant from February to April in the plains of Uttar Pradesh. After that most of them die due to heat and only few survive in moist places.

Control

1. Since bugs congregate on the leaves and stems, they may be collected and destroyed.
2. Irrigate the field by mixing 5 kg of crude oil emulsion per hectare to destroy the hiding bugs in cracks and crevices.
3. Dusting of crop with following insecticides have been found quite effective :
   (a) Malathion 5 % dust @ 25 kg/hect.
   (b) Carbaryl 10% dust @ 25 kg/hect.

4. Spraying of any one of the following insecticides is also effective to control the bug :
   (a) Malathion 50 Ec @ 1.25 lit/hect.
   (b) Dimecron 85% @ 0.250 lit/hect.
   (c) Rogor 30 EC @ 1.00 lit/hect.

Natural enemies : The following parasites have been found on different stages of the host.

Egg Parasite

(i) *Liophanunus samueli*

(ii) *Typhodytes sp.*

Nymph and adult parasite : *Alphora sp.*

Til Leaf and Pod Caterpillar

Scientific name : *Antigastra catalaunalis* Dup.

Order : Lepidoptera

Family : Pyralididae

Host plant : Til and wild varieties of *Sesamum* genus.

Distribution : The pest is found in Europe, Africa, Cyprus, Malta, Indonesia and South-East Asia. In India it is widely distributed in sesamum growing areas.

Nature of damage : The damage is caused by the caterpillars which feed on apical shoots and young pods. The young caterpillar feeds on the leaves. They bind together the tender leaves of the growing shoot with the help of silken threads produced by them and prefer to remain within the bunch foliage thus rolled and spun together in a sort of webbed mass.

The size of the bunch become bigger and bigger as the larva grows in size and incorporates more and more of the neighbouring foliage. They also bore into the shoots, flowers buds and pods. An early attack may destroy the whole plant and single larva can destroy 2-3 young plants.

Life history : The four developmental stages *viz.*, egg, larva, pupa and adult are found in its life cycle which have been described below :

Egg : The female moths lay eggs singly generally on the underside of leaves and on the apical portions of the growing shoot: The egg are greenish or shining pale-green in colour. A single female can lay upto 150 eggs which hatch in 2-7 days depending on the season.

Larva : The newly hatched larva is 2 mm in length and some what whitish in colour; later on the colour slowly changes to green and develop black dots all over the body. The full grown larva measures 7 to 15 mm in length. The larval period lasts for 10-33 days depending upon environmental conditions.

Pupa : The full grown larvae creep to the ground and pupate in silken cocoons in the soil. Sometimes pupation takes place on the plant itself. The pupa is greenish when just formed but changes to brown later on. The pupal period lasts for 4-20 days depending upon the season.

Adult : The adult is a medium sized moth with a wing span of 15-20 cm. The fore wings are reddish-yellow in colour and there are zigzag distinct reddish decorative markings on them. The hind wings are pale-yellow or transparent. They are nocturnal in habit and remain hide during the day under fallen leaves. The moths are phototropic and are attracted to light. They live about 3 weeks and starts egg laying within 4-5 days of emergence.

The life cycle is completed in 19-64 days and there are about 14 generations in a year. All stages of the insect are seen in the autumn or even upto December. During January and February, the insect hibernates as a larva inside the stable or the harvested stalks.

Control

1. Early showing of the crop minimise its damage.
2. The infested pods and webbed leaves may be collected and burnt.
3. Spraying the crop with anyone of the following insecticides :

   (i) Chloropyrith — 0.05%

   (ii) Quinalphos — 0.05%

   (iii) Carbaryl — 0.15%

   (iv) Endosulfan — 0.07%

Hawk Moth

Scientific name : *Acherontia styx* Westwood

Order : Lepidoptera

Family : Sphingidae

Host plants : Til, potato, brinjal, sem and balsam etc.

Distribution : It is common in India and has been described a pest of til and is also reported from Indonesia, Sri Lanka, Philippines and Mayanmar.

Nature of damage : The damage is caused by the caterpillar, but moth is also harmful as it sucks honey from the honey combs in apiaries. The caterpillars feed voraciously on leaves and defoliate the plants. During heavy infestation all the leaves of plants are eaten-up by the caterpillars and leaving all the stalks.

Life history : The four developmental stages namely egg, larva, pupa and adult are found in its life cycle which have been described below :

Egg : The female moths lay globular eggs singly generally on the under-surface of the leaves of the food plant and the eggs are fairly large. Their colour is at first greenish-white but later on they turn yellow. The incubation period varies from 2-5 days.

Larva : The newly hatched larva is pale-yellowish in colour. It begins to feed on the leaves and develops broad green stripes running obliquely along the dorso-lateral surface. It is a typical sphingid caterpillar with the characteristic head and caudal horn. When the larva is full grown, its colour undergoes a rather drastic change to brown.

The full grown caterpillar is about 5 cm in length and 1 cm in width, often retracts some of its anterior body segments and looks like a sphinx. The horn like projections on the hind end of the abdomen are conspicuous. The body of the caterpillar is plump and is decorated with a pleasant mixture of soft colours. The larva looks quite ferocious and venomous, although it is quite harmless and safe to handle. The larval period lasts for about 2 months.

Pupa : The full grown larva descends from the plant to the ground and finds out a suitable place for entry into the soil wherein it pupates. The pupal period lasts for 2-3 weeks.

Adult : The adult is a large reddish-brown moth with a wing span of about 10 cm. The fore wings are decorated with a mixture of dark brown and grey patterns with dark or black wavy markings and a prominent yellow spot on each wing. The hind wings have an ochre background on which there are two broad dark brown wavy cross stripes. There is a prominent Death's head mark on the thorax. They are swift fliers and often make hawk-like darts to a light source soon after dusk.

The life cycle is completed in 78 to 89 days and there are 3 generations in a year. The pest is active throughout year except winter, which is passed in pupal stage in soil. The change in colouration of the larvae and adults aid them in protective mimicry.

Control

1. Hand picking in small areas is quite effective for such large-sized insects.
2. Dusting the crop with 10% carbaryl dust @ 25-30 kg/hect.
3. Sparying the crop with 0.07% endosulfan or 0.15% carbaryl @ 750 lit/hect.

Linseed Gall-midge

Scientific name : *Dasyneura lini* Barnes

Order : Diptera

Family : Cecidomyidae

Host plants : Linseed.

Distribution : It is distributed in different states of India including Andhra Pradesh, Madhya Pradesh, Bihar, Uttar Pradesh, Delhi and Punjab.

Nature of damage : The damage is caused by the maggots, which feed on the flower buds and prevent their proper opening. Consequently the seed does not set properly. The corolla of the infested flowers get crumpled and their essential organs look emaciated even of the flowers open. Due to their feeding, galls are produced and there is no pod formation. The incidence of their pest goes up to 20 per cent.

Life history : Four developmental stages, *viz.*, egg, larva, pupa and adult are found in its life cycle, which have been described below :

Egg : The female fly lays smooth, transparent eggs in the folds of flowers or in tender green buds. A female lays 29-103 eggs, either singly or in clusters of 3-5. The hatching period of eggs varies from 2 to 5 days. The length of the eggs varies from .22 to .28 mm and width range between 0.04 to 0.08 mm.

Larva : The newly hatched maggots are transparent, with a yellow patch on the abdomen. The maggots feed inside the flower-buds and eat the contents. The full grown larva becomes deep pink in colour and measures 2 mm in length. They pass through 4 instars and become full grown in 4-10 days.

Pupa : The full-grown maggots drop to the ground, prepare a cocoon and pupate in the soil. The pupal period is about 4-9 days. The female pupa is 1.9 mm long and 0.64 mm wide, while the male is smaller being 1.6 mm long and 0.5 mm wide. It is dark orange in colour.

Adult : The linseed gall-midge, is a minute (male slightly over 1 mm and female about 1.5 mm in length) and of beautiful orange colour. Thorax is dark dorsally and rest of the thorax and abdomen fuscous.

The linseed gall-midge appears in December or early in January. Its appearance is correlated with the appearance of the blossoms of linseed. The total life cycle is completed in 11-26 days and there are 4 overlapping generations in a year. The midges continue to emerge till the end of February or beginning of March depending on humidity and temperature.

Control

1. The crop should be sown easily preferable in the 1st fortnight of October.
2. The crop may be sprayed with anyone of following insecticides twice, firstly at the time of initiation on bud formation and secondly after 15 days of 1st spray:
   (i) Phosphamidon : 85 SL : 250 ml.
   (ii) Monocrotophs : 36 SL : 750 ml.
   (iii) Methyl-o-demeton : 25 EC : 1.0 lit.
   (iv) Endosulfan : 35 EC : 1.5 lit.

Natural Enemy

Systasis dasyneura : parasite on maggots.

Cotton Leaf Roller

Common name : Ban ki surhi, Patti mor surhi

Scientific name : *Sylepta derogata* Fabr.

Order : Lepidoptera

Family : Pyralididae

Host plants : Cotton, bhindi, sunnhemp, hollyhock and other plants of Malvaceae family.

Distribution : Besides India, it is found in Pakistan, Mayanmar, Sri Lanka, China, Japan, Jawa, Egypt, Africa and Australia etc. In India it is distributed in all the cotton growing areas, specially Gujarat, Andhra Pradesh, Bengal, Maharashtra. Punjab and Uttar Pradesh.

Nature of damage : The damage is done by the caterpillar. The newly hatched larva feeds on the lower surface of the leaves mainly epidermis but when it grows older rolls up the leaves in to a funnel shape and feeds from the margin. The larva remains hidden beneath the rolled leaves. Initially they are found more in number on single leaf (more than 30) but later only one is found on a leaf. Some times it appears in epidemic form and in such case 3,000 larvae have been found in 100 sq. feet (*Nag Pal,* 1948). Practically, the whole leaf is eaten up or big holes are made with the result it drops to the ground. The crop sown on the onset of monsoon and provided with nitrogenous fertilizers are subjected to much more attack of the insect. As a result of leaf damage, the photosynthesis, respiration and transpiration activities are badly affected due to which growth of the plant is affected and ultimately bears lesser bolls. The American cotton is more preferred than the Desi cotton.

Life cycle **:** There are four developmental stages, *viz.*, egg, larva, pupa and adult in its life cycle which are described below.

Egg **:** The female moth lays eggs singly on the under side of the leaf along the thicker veins during evening hours or at night. The eggs are flat oval and white in colour which turn to brown at the time of hatching. Single female lays about 250 to 300 eggs in her life span. The hatching time varies from 2-9 days while pre-oviposition period is 2-3 days.

Larva **:** The young larva feeds on the lower side of the leaf and spin threads over and around it self as a protection. Later, it turns the edge of the leaf over in a fold and binds it down with silken thread, living safely within this fold. The full grown larva is green in colour with dark brown head and bears a brown coloured notch in the prothorax. At the time of pupation it becomes pinkish. It moults four times and become full grown in 15 to 35 days.

Pupa **:** The pupation takes place either in rolled leaves or on the ground among fallen leaves. Before changing in to pupa, the larva shrinks and become pinkish in colour. The pupa is reddish brown in colour and measures 12 mm. Pupal period lasts in about 12 days.

Adult **:** The adult moth is whitish in colour with faint yellow tinge, the wings with many fine dark lines forming an irregular pattern. It measures 12 mm long with wing expanse of 25 mm. Labial and maxillary palps are fused together to form a snout.

The pest over winters over as a full grown pinkish caterpillar among the fallen leaves. These larvae pupate in the second week of February, from which moths appear in March. The pest is active from March to October-November. There are 5-6 generations in a year but in Bengal only 3 generations are found. The total life cycle is completed in 25 to 54 days depending upon environmental conditions.

Control

1. The field should be ploughed deeply before sowing so that hibernating larvae may be destroyed.
2. Regular hand picking of rolled leaves minimise the attack of pest.
3. Balance nitrogen should be given and if possible resistant varieties, *viz.*, 10 F should be preferred.
4. The crop may be dusted with any one of the following insecticides to control the pest :
   (a) Carbaryl 10% dust @ 20 kg/hect.
   (b) Fenitrothion 5% dust @ 20 kg/hect.
5. Effective control can also be achieved by spraying the crop with any one of the following insecticides :
   (a) Carbaryl 50 WP-3.00 kg/hect.
   (b) Endosulfan 35 EC-2.50 lit/hect.
   (c) Formothion 25 EC@ 1.00 lit/hect.
6. Spray of 0.1 percent brestenol 45 WP protects the crop. It acts as antifeedant.

Natural enemies : The following are the effective natural enemies of the pest :

Egg parasite : *Trichogramma minutum, R.*

Larval Parasite

(a) *Bessa remota,* Alder.

(b) *Microbracon lefroyi* D.

(c) *Brachymeria tachardiae,* Cam.

(d) *Xanthopimpla punetata,* F.

(e) *Microtoridea lissonara,* Vier.

(f) *Elasmus indicus,* Rohw.

Pupal parasite : *Xanthopimpla sp.*

Many birds also prey on the larva of the pest which may be encouraged.

Spotted Boll Worm

Common name : Chitkabri surhi

Scientific Name

(i) *Earias vitella (fabia),* Stoll.

(ii) *Earias insulana,* Boisd

Order : Lepidoptera

Family : Cymbidae

Host plants : Cotton, Bhindi, Sunnhemp, Hollyhock and other Malvaceous plants.

Distribution : It is found in Spain, Syria, Palestine, Egypt, India, Mayanmar, Madagascar and Australia. In India it is commonly occur in M.P., Maharashtra, Mysore, Chennai, Gujarat, U.P., Delhi and Punjab etc.

Nature of damage : The larva is the destructive stage of the pest which in the beginning (when plants are small 20-30 cm high) bores the top portion of the growing shoot which withers and drops.

As a result of which plant growth is adversely affected. *Lefroy* (1906) stated that this pest is more serious than locusts. When the flower buds and bolls appear the larvae start feeding on them. The caterpillars bore in to bolls close the hole with their excreta and feed inside.

The attacked bolls are generally shed those left on the plants open prematurely with the result poor quality of fibre is produced and market value is reduced, Several holes plugged with excreta can be seen on the bolls and a single larva may destroy many bolls in its life time. The damage caused by the pest is estimated to the tune of 20-40% by Claiston (1921) and 50% by Mishra (1938).

Life history : There are four developmental stages namely egg, larva, pupa and adult in its life history which are described as follows :

Egg : The female moth lays eggs generally singly, scattered over fresh squares etc. The eggs are spherical, bluish green in colour with parallel longitudinal ridges which projects upward and give it crowned appearance. The female lays 200-500 eggs in her life time only at night. An individual egg is 1 mm in diameter. The hatching period of eggs varies from 3-4 days in summer and 8-9 days in winter.

Larva : The newly hatched larva is brownish, white with a dark head and prothoracic shield. It measures 1.2 to 1.3 mm in length. The caterpillar remains about 24 hours outside the shoots or bolls and after that it enters in to the shoot or boll. Larva moults 3 or 4 times become full grown measuring 22-25 mm long.

The caterpillar of *E. vitella* is brownish with a median longitudinal streak while of *E. insulana* has dull greenish body with a number of black marks and oval, orange dots on the prothorax. The caterpillar becomes full grown in 10-12 days during summer and 15-20 days during winter.

Pupa : The full grown caterpillar comes out of the bolls and pupates either on the plant or on the ground among fallen leaves or 5-25 cm deep in soil. The pupa is enclosed in tough silken cocoon of dirty white or light brown colour. It measures 12 mm in length and 4-5 mm in breadth. Pupal period lasts for 5-15 days during summer and rainy season and 18-31 days during winter.

Adult

Earias vitella : The moth measures 12 mm long and 21 across the wings, the fore wings have a broad green band extending from the base to apex and hind wings are white in colour. The body is pinkish white in colour having tuft of hairs behind the thorax.

Earias insulana : It measures about the same as above, but fore wings are completely green. The species is more prevalent in dry regions.

The pest remains active throughout the year, but population is low during December to March. It begins to multiply in April-May and reaches its maximum during July to September. The life cycle is completed in 31-55 days and 7-8 generations are found in a year. The males die after copulation while females survive 10 to 20 days depending upon the season.

Control

1. Destruction of plants which harbour the pest.
2. After harvesting of crops the field should be ploughed to destroy the hidden caterpillars.
3. Collection and destruction of infested shoots and bolls etc.
4. The crop may be sprayed with any one of the following insecticides to minimise the attack of pest :
   (a) Thiometon 25 EC @ 1.25 lit/hect.
   (b) Endosulfan 35 EC @ 2.5 lit/hect.
   (c) Fenetrothion 50 EC @ 3.00 lit/hect.
   (d) Phosphomidan 85% @ 500 EC lit/hect.
   (e) Carbaryl 50 WP @ 2.5 kg/hect.

The crop should be sprayed twice first at the time of hatching of the eggs (peak hatching period) and second after

15 days of previous spraying. The quantity of water may be required 700-750 lits/hect. If the insecticides are to be applied on bhindi crop endosulfan or D D V P should be preferred.

Natural enemies : The following are the natural enemies on different stages of the pest :

Egg parasite : *Trichogramma spp.*

Larval Parasite

(a) *Microbracon lefroyi, D* and *G*

(b) *M. Hebets, Say*

(c) *Elasmus sp.*

Pupal Parasite

(a) *Melcha mursei,* Carn.

(b) *Chelonus rufus,* Lyle.

(c) *Chalchis* sp.

Pink Boll Worm

Local name : Kapas ki Lal Surhi

Scientific name : *Pectinopliora (Platyedra)* gossypiella Saund.

Order : Lepidoptera

Family : Gelechiidae

Host plants : Mainly cotton, but some time on Deccan Hemp, Kanghi and Gulkhara etc.

Distribution : It is widely distributed and probably the most serious cotton pest on a world basis. Besides India, it is mainly found in Mayanmar, Sri Lanka, Egypt, South Africa, Maxico, Brazil, Hawaii, Philippines and Australia. In India it was first of all found in Mumbai (Maharashtra) during 1842 and now distributed in U.P., Punjab, Rajasthan, Andhra Pradesh, Gujarat, Bihar and other cotton growing states.

Nature of damage : The damage is caused by the larvae which bore in to flower buds, flowers and young bolls. The young attacked bolls are invariably fall down. The pink boll worm larvae do most spectacular damage to practically mature cotton bolls in which they enter at such a tiny stage (just-hatched larvae) that their entry hole gets healed and in which they remain devouring both seed and fibre-forming tissues.

The infestation some times is so severe that up to 10 caterpillars may be found in a single boll and 75 to 100 per cent bolls may be found infested. The oil content and the germinating capacity of the seed and lint formation are badly affected. The American cotton is more damaged than the Desi cotton and may cause 25-50% loss.

The larva of spotted boll worm prefer to eat the gynaecium while pink boll worm eats the anthers of the flowers, the holes on big bolls are plugged by the former while remain open in case of later.

Life history : There are four developmental stages, *viz.,* egg, larva, pupa and adult are found in its life cycle which are described as below :

Egg : The female moth starts egg laying after 1 or 2 days of copulation. The eggs are laid singly or in batches of 2-10 on leaves, buds, flowers and bolls. A single female can lay 125 to 300 eggs which are 0.50 mm in length and 0.30 mm in breadth. The eggs are flattened, elongated and nearly white when laid freshly, becoming brown on the third day and deep brown at the time of hatching. The hatching period of eggs varies from 4 to 29 days depending upon the environmental conditions in U.P. during rainy season, it hatches in 4-5 days.

Larva : The newly hatched larva is colourless and about 1 mm in length. It bores in to the boll, begins to feed one and when one seed is finished, it passes on to the next. The entry holes get quickly healed up and it is difficult to distinguish an infested boll from an uninfested one. The full grown caterpillar is 13 x 2 mm long and pinkish in colour. The larvae undergo generally 3 moults and the larval period depends on two distinct types of life cycles referred as short life cycle and long life cycle. The short life cycle is completed just like other lepidopterous pest in 30 to 90 days.

In case of long life cycle, the larvae enter in to resting stage where they remain over a period of 8-12 months and some times up to 2 years before they pupate and moths emerge. The interesting point in the long life cycle is that at certain times of the year two types of larvae are found in a mixed population. One type pupates in the normal manner, while the other type makes preparations for prolonged hibernation.

The larva scoopes and binds together around it self generally two but some time more than two seeds with the help of silken threads and then enters the resting stage in the hibernaculum prepared between double seeds. It is this stage which has been responsible for the universal distribution of the pink boll worm in the cotton growing regions of the world. When this double-seed is sown along with healthy seeds, the resting larva comes out of its hibernaculum and makes a tunnel up to the surface and then goes down to pupate.

Pupa : The pupation takes place in a silken cocoon on the ground among fallen leaves and flowers etc. The pupa is light brown in colour and measures 6 mm in length and 2.6 mm in breadth. At the time of hatching the colour of pupa changes in to dark brown. The pupal period lasts in 6-20 days.

Adult : The adults are small moths of dark, fuscous brown colour and about 1 cm in body length. The antennae are filiform, the palpi long and curved. The fore wings are pointing towards margins and possesses long brown fringes. They are active during the night.

The moths are comparatively quite long-lived. Male generally survives up to 20 days, while female up to 55 days. Under normal conditions, the life cycle is completed in 22 to 77 days but the long-life cycle is completed in about 13.5 months in U.P. There are 4 to 6 generations in a year.

Control

1. Collection and destruction of infested flowers, flower buds and bolls etc.
2. Heat all cotton seeds at about 104-110F., either in the sun or by simon cotton seed heater to kill the hibernating larvae in them.
3. Seeds should be fumigated, either with methyl bromide @ 3.5 kg/100 sq. m ṣeeds for 24 hrs or with aluminium phosphide @ 45 tablets/100 sq. m seeds for 7 days before sowing.
4. Ratooning of cotton should not be practiced in infested areas and all other alternate hosts should be destroyed.
5. The spraying may be done with any one of the following insecticides at fortnightly intervals twice or thrice to destroy the pest :

   (a) Thiometon 25 EC @ 1.000 lit/hect.

   (b) Phosphomidon 85% @ 0.500 lit/hect.

   (c) Endosulfan 35 EC @ 2.500 lit/hect.

   (d) Fenitrothion 50 EC @ 3.000 lit/hect.

   (e) Carbaryl 50 WP @ 2.500 kg/hect.

Natural enemies : The following are natural enemies found in different stages of the pest.

Egg Parasite

(a) *Triphlapi pectinophorae*

Larval Parasite

(a) *Microbracon lefroyi* (D and G)

(b) *M. greeni*

(c) *Bracon* kitchneri (Will)

(d) *Apanteles pectinophorae*

(e) *Elasmus pus ptyedrae* Ferr.

Legal control : As per direction of Director of Plant Protection, Quarantine and Storage, Ministry of the Food and Agriculture India, all imported cotton must be fumigated at ports.

Gram Cut Worm

Common name : Kata, Kirauna, Kamta, Katuwa, Surhi

Scientific name : *Agrotis ypsilon* Rott.

Order : Lepidoptera

Family : Noctuidae.

The other two spp. are as follows :

Agrvtis flammatra Schiff.

A. segetum Schiff.

Host plants : Gram, Potato, Pea, Wheat, Tobacco, Cotton, Mustard, Groundnut, Bhindi, Lucerne, Cabbage etc. Thus they are polyphagous in nature and attack many crops.

Distribution : Besides India, it is distributed in America, Africa, Sri Lanka, Mayanmar, New Zealand, Malaya and European countries. In India the cut worms are found in all the states with particular reference to U.P., Bihar, Rajasthan and Madhya Pradesh.

Nature of damage : In India winter is generally lean season for insect-pest activity, but the cutworms belong to that small group of insects, the destructive activity of which is more marked in the rabi crops. They are known as cutworms because these cut the plants and fell down to the ground either the whole plants like gram or their twigs. They are some times also called as surface caterpillars because their activities are mostly confined few centimetres of the surface soil. The damage is caused by the larvae which hide during the day and come out in the evening to cause the damage. First of all they feed on the epidermis of the fallen leaves or the green leaves touching the ground. Later, the caterpillars cut the leaf or shoot or the plants just above the ground level and burried in the soil. They destroy many more plants than they actually feed upon and thus cause a very serious loss. The attack of cut worm in gram is confined from November to February.

Life history : The four developmental stages, viz., egg, larva, pupa and adult are found in its cycle which are described as below :

Egg : The moths of the cut worms are strong fliers. They come to the plains in winter from hills and return in summer to hills again. After mating, the females lay eggs mostly on the underside of the leaves or on the soil near the root base. The eggs are laid singly or in clusters of up to 30 eggs. A female generally lay 300-350 eggs but can lay thousands of eggs. The eggs are round yellowish in colour and measuring 1 mm long. The hatching period of eggs varies from 2-6 days but during severe cold 12-15 days.

Larva : The young larvae feed on their egg-shell as their first meal and fall to the ground at the slightest disturbance. They are light bluish-green in colour and measuring 2 mm long. The caterpillars remain hidden in the loose soil during the day and come out at night. The full grown caterpillar is about 40-45 mm in length, plump, smooth and flattened in look and dull grey brown in colour. The caterpillar becomes full grown in 21-36 days.

Pupa : The pupation takes place in the soil. The full grown caterpillar enters the soil much deeper than during its usual activity and constructs the earthen chamber, the sides of which are made smooth and hard by an exudation from the larval mouth. The pupae are reddish brown in colour and about 20 mm in length and 7 mm in breadth. The pupal stage lasts for 10-30 days.

Adult

Agrotis ypsilon : The moths are brown in colour measuring 25 mm in length and 7 mm in breadth with a wing expanse of 40-50 mm. The males have bi-pectinate antennae while females have filiform. Fore wings are dirty white in colour having dirty brown spots while hind wings are white. The terminal end of abdomen bears a tuft of hairs.

Agrotis flammatra : The fore wings are characteristically marked and two third of costal area is pale, grey brown having a kidney shaped and a semicircular spot below the pale area and two black wavy lines near the base. A black sign of inverted V (^) is found on prothorax.

Agrotis segetum : *It* is mostly found in hills and smaller than the above two species. The colour of the body is brownish-orange.

They are active from October to March in plains and after which most of them migrate to hills. The life cycle is completed in 35 to 84 days and generally two generations are found in plains and in the hills four generations occur during summer.

Control

1. The moth may be destroyed after attracting them in light-traps.
2. Hand-picking of larvae may also reduce the population of caterpillars.
3. Flooding of field may kill the caterpillars.

4. Any one of the following dusts may be mixed in the soil at the time of sowing to kill the larvae of the pest:
   (a) Methyl parathion 2% dust @ 25 kg/hect.
   (b) Heptachlor 3% dust @ 35 kg/hect.

   BHC and aldrin dusts should not be used in tuber crops.
5. Any one of the following insecticides may be sprayed near the base of plants in standing crop to control the pest :
   (a) Quinalphos 25 EC @ 3.5 lit/hect.
   (b) Gamma BHC 20 EC @ 3.75 lit/hect.
   (c) Chlorpyriphos 20 EC @ 3.5 lit/hect.

Natural enemies : The following insects are the larval parasite of the pest :

(a) *Braconis* wasp

(b) *Apanteles* sp.

(c) *Microbracon* sp.

Gram Caterpillar

Common name : Chane ke illi or surhi, Fali bedhak.

Scientific name : *Helicoverpa (Heliothis) armigera* Hubn.

Order : Lepidoptera

Family : Noctuidae

Host plants : Gram, Pea, Cotton, Arhar, Maize, Cobs, Tomato and Tobacco etc.

Distribution : It is widely distributed and practically omnivorous pest. In United States of America, it is a serious pest of cotton and is known as the cotton boll worm. In India it is mostly found in Punjab, U.P., Rajasthan, Bihar and Madhya Pradesh.

Nature of damage : The pest is active through out the year but damage to gram is cause from November to March. The damage is caused by the larvae which feed on the leaves and destroy the seedlings in the early stages.

At the time of pod formation it is found feeding on the developing grain after cutting a hole in the pod and thrusting its heads therein. It does not hide in the soil during day but remains hide in the plants. It has been estimated that a single caterpillar destroys 30-40 pods of gram in its life time. It has been observed that grown caterpillars feed on the small larvae of its own which is known as cannibelism. In severe infestation damage may be caused from 20-50 per cent.

After the harvest of gram crop in March-April it migrates to tomato and feeds on ripening fruits which invites the roting of the fruits.

Life history : Four developmental stages, *viz.*, egg, larva, pupa and adult are found in its life cycle which are described as below :

Egg : The adults moths are visible in the gram field during early part of November which are active after dusk. Two to four days of copulation female lays light yellow eggs generally on the side of the leaves or on twigs or on pods. The eggs are spherical, sculptured and with greenish shine. A single female can lay 500 to 1000 eggs in her life span. The egg-period lasts from 3 to 7 days.

Larva : The newly hatched larva measures 2 to 2.5 mm long and light-green in colour. The young larva begins to feed on the tender portions of the leaves and shoots and when pod formation takes place it feeds on the pods. The full grown caterpillar is about 35 mm in length and very variable in colour. Generally it is pale green with a dark mid-dorsal and two yellow lateral stripes. The caterpillar moults four times and becomes full grown in 20-25 days.

Pupa : The larva when full grown drops to the ground and pupate in a shell 8 to 10 cm deep in the soil. The pupa is dark brown in colour and about 16 mm in length. It remains in the soil during the cold months of January and February. The pupal period generally lasts from 10 to 24 days depending upon the environmental conditions.

Adult : It is a typical stout-built noctuid moth with a wing span of 30-40 mm, and colour varying from olive grey to red brown. The fore wing has dark specks and dark area near the outer margin. The hind wings are white with dark veins and with a broad blackish band along the outer margin. There are filiform antennae in males and bipectinate in females. The moths of *H. armigera* are 20-25 mm in length while that of *H. asulta,* 16 mm in length with a wing span of 37 mm.

The moths emerge in late February or early March from over wintering pupae and lay eggs again. The total life cycle is completed in 35-70 days on gram.

Control

1. Deep ploughing of infested field in the summer to expose the pupae is found useful.
2. The adult moths may be attracted in light-traps and destroyed. Any one of following dusts may be used to destroy the pest :
   (a) Malathion 5% dust @ 25 kg/hect.
   (b) Carbaryl 10% dust @ 25 kg/hect.
   (c) Fenitrothion 5% dust @ 20 kg/hect.
   (d) Phenthoate 2% dust @ 25 kg/hect.
4. Spraying with any one of the following insecticides has also been found effective to control the pest :
   (a) Endosulfan 35 EC @ 1.25 lit/hect.
   (b) Carbaryl 50 WP @ 2.50 kg/hect.

(c) Gamma BHC 20 EC @ 2.5 kg/hect.

(d) Monocrotophos 36 SC @ 1.00 lit/hect.

The above quantity of insecticide may be mixed in 700-750 tit of water and sprayed on the crop.

Natural enemies : The following are the important natural enemies :

Egg Parasite

(a) *Trichogramma minutum*

(b) *Triphleps insidious*

Larval Parasite

(a) *Apanteles sp.*

(b) *Bracon sp.*

(c) *Chelonus narayani*

Pea Leaf Miner

Scientific name : *Phytomyza atricomis* Meigen

Order : Diptera

Family : Agromyzidae

Host plants : It is polyphagous pest mostly feeds on pea, crucifers, Unseed and potato.

Distribution : Besides India, it is found in Europe, America, Russia, Australia and New Zealand etc. In India, it has been recorded from 29 different species of economic plants belonging to 13 families from different states.

Nature of damage : The larvae are the damaging stage which make prominent whitish tunnels in the leaves. If the attacked leaves are held against bright light, minute slender larvae an be seen feeding within the tunnels. The huge number of tunnels made by the larva interfere the photosynthesis and the proper growth of plants. The female fly also punctures the tender leaves with its ovipositor and then turns round to feed on the plant juice exuding from the puncture.

Each fly makes numerous such punctures of each leaf and these injured points develop into as many protuberances of the leaf *Saxena* and *others* (1969) recorded a severe infestation of the leaf minor ranging upto 85.7 per cent.

Life history : The four developmental stages, *viz.*, egg, larva, pupa and adult are found in its life cycle which are described below :

Egg : The flies emerge in the beginning of December and after mating start egg laying singly in the leaf tissues. The eggs are very small, about 0.3 mm in length, and oval elongate in shape. A female fly lays 350-400 eggs and the incubation period is 3 to 7 days.

Larva : The tiny maggots after hatching from the eggs begin to feed on the mesophyl of the leaf without damaging the two epidermal layers. Maggots remain within the tunneled gallaries and become full grown in 4-10 days. The full grown larva is about 3 mm long and light pinkish in colour.

Pupa : The pupation takes place within the gallaries where larva pupates within its last larval shin which hardens in due course. The pupa becomes reddish-brown or dark-brown in colour. The pupal period ranges from 6 to 12 days.

Adult : The adult is very small fly, about 1.5 mm in length and with a wing span of 3.5 mm. It is two winged, which has a greyish-black mesonotum and yellowish frons.

The pest is active from December to May and from May onwards it passes the rest of the year in soil in the pupal stage. The total life cycle is completed in 14-21 days and 4-5 generations are generally found in a year.

Control

1. Collect the damaged portions of the plant during initial stage and destroy to minimise the multiplication.

   The pest can be controlled by repeated spraying (after every 2-3 weeks) of anyone of the following insecticides:

   (i) Dimethoate : 0.03%

   (ii) Formothion : 0.03%

   (iii) Carbaryl : 0.1%

   (iv) Monocrotophos : 0.03%

Pea Bod Borer

Scientific name : *Etiella zinckenella*, Treitscke

Order : Lepidoptera

Family : Phycitidae

Host plants : Pea, lentil, red gram, lathyrus and sunnhemp.

Distribution : It is distributed throughout India with particular reference to Uttar Pradesh, Bihar, Madhya Pradesh and Punjab.

Nature of damage : Newly hatched larvae constructs a web type structure through which they entered the pod. After 2-3 days of feeding inside the seed, the caterpillar come-out to feed superficially on the seeds. They migrate from one seed to another inside the pod without completely consuming the damaged seeds. In the later stage of damage, the pods become discoloured, watery in appearance and emitted foul smell. On an average 77.6 per cent of seeds in infested pods are found damaged by the pest. The yield of pea-seeds is reduced on an average by 23.9 per cent due to the feeding of the larvae.

Life history : The four developmental stages namely egg, larva, pupa and adult are found in its life cycle which are described below :

Egg : The eggs are laid on blossoms, buds, twigs and leaves of the plants. They are generally deposited singly but sometimes also in batches of 4-6 eggs. A single female lays 44 to 178 eggs in her life span. The incubation period varies from 4 to 7 days which may extend up 33 days in adverse conditions.

Larva : The newly hatched caterpillar is yellowish in colour which changes to milky white after 10-12 hours of feeding. The colour further changes to light green after 6-7 days of feeding and ultimately it becomes pinkish when full-fed. The larva bores into the pod and feeds on the grain inside. Often more than one larva is observed in a pod. The larval period lasts for 10-17 days. Cannibalism has also been observed amongst young and between young and older caterpillars..

Pupa : Full grown larva comesout of the pod and pupates in soil. Pupation takes place in cocoons in the soil at a depth of 2 to 4 cm. Pre-pupal stage ranges from 2-4 days. Pupal period lasts for 10-20 days depending upon the temperature.

Adult : The emergence of moth takes place at night and in a few cases in the early hours of the morning. The moths are grey in colour with wing expanse of 25 mm. The fore wings have dark marginal lines and are interspersed with ochreous scales. The moths emerge in February and March nocturnal in habit.

The pest breeds throughout the year. The total life cycle is completed in 27 to 46 days and passes through 5 generations. The moths start appearing in the fields in during last week of February. All the stages of the pest are available in the field during the months of March and April.

Control

1. Spraying the crop with anyone of the following insecticides per hectare :

   (i) Methyl demeton 25 EC : 1.50 litres.

   (ii) Endosulfan 35 EC : 1.5 litres.

   (iii) Carbaryl 50 WP : 2.25 kg.

2. Dusting of any one of the following insecticides has also found effective :

   (i) Endosulfan 2% dust : 25 kg.

   (ii) Carbaryl 5% dust : 25 kg.

   (iii) Fenitrothion 5% dust : 25 kg.

5

Fruit Destructors

Mango Mealy Bug

Local name : Safed phunga, Gujia, Aam ka matkun.

Scientific name : Drosicha mangiferae (stebbingi) Green.

Order : Hemiptera

Sub-order : Homoptera

Family : Coccidae

Host plants : The list of food plants includes about 62 species of trees, shrubs and herbs with particular reference to mango, guava, peaches, plump, rose, castor, jack fruit, papaya, fig and lemon etc.

Distribution : The pest is distributed in India, Pakistan, Sri Lanka, Mayanmar, Malaya, China and Formosa etc. In India, it is mainly found in Assam, Bihar, Delhi, Chennai, Orissa, Punjab, Uttar Pradesh and Madhya Pradesh.

Nature of damage : The damage is caused by the nymphs of both sexes and female adults which suck the sap from twigs, shoots and flowers. As a result of which flowers dry up and only few fruits are set. The fruit bearing twigs are so weak that with the slightest jerk whether by bird or wind, the fruits fall down. They secrete honey-dew which attracts fungus due to which black spots may be seen on the twigs and shoots etc. The attack is much more serious during spring when shoots are richly supplied with overflowering cell sap. The attack persists from November to April-May and by that time it changes in to adult. The damage caused by this pest is estimated to the tune of 20-50 per cent depending upon the severity of attack.

Life circle : The female and male adults develop differently. In the case of female 3 developmental stages, *viz.*, egg, nymph and adult are found, while in case of male egg, nymph, pupa and adult are found.

Egg : The adult bugs are common from the end of March to the middle of May. The adult gravid females after fertilization crawl down along the tree trunk to the ground or some times just fall down. It lays eggs in the soil at a depth of about 8-15 cm in clusters of about 300-400 eggs which are enclosed in a sac-like pouch of some silken material, after 15-16 days of copulation. The eggs in such a cluster are deposited by the female in instalments over a period of 7-15 days. The oviposition is generally confined to an area of a few feet in diameter round the base of the tree. The eggs are cylindrical pinkish-brown in colour and measuring 1.25 x 1.00 mm long. The eggs laid generally during April and May remain in the soil till the end of November or early December, when they start hatching.

Nymph : The newly hatched nymphs are red in colour and may be seen crawling up the tree trunk. The 1st instar nymphs of both the sexes can not be differentiated. The 1st instar nymphs are 1 mm long and moult in the month of January about two months after hatching. The IInd instar nymphs on the basis of length may be differentiated; female nymphs being larger of the male. The development of both the, sexes takes place as follows :

Development of female nymph : The second instar nymph moults in the month of March about 1 and 1/2 months of 1st moult and measures 8 mm long at this stage. Initially they are brown in colour and after some times they are covered with white powder resulting in to white in colour. They grow gradually and become 1 cm long. In the mean time reproductive organs are developed. Some times they moult three times, then reproductive organs developed in fourth instar.

Development of male nymph : In the case of males, after first moulting a short prepupal and pupal stages are noticed. The pupation takes place in the soil on the fallen leaves in cocoons made by waxy thread secreted by nymphs. Pupal period is of about 21 days.

Adult : There is a well established sexual dimorphism in this stage which is found from April to June.

Male : The male is the winged insect with only one pair of wings and a very delicate reddish body. It measures 5 mm x 2 mm in length and breadth, respectively with a wing span of 13 mm. The antennae are whorled type and black in colour. They live about a week and having no mouth parts hence do not feed.

Female : The female is wingless fleshy flat bodies insect with a length of 14-16 mm and a breadth of 8.5 mm. Its body is covered with ashed white mealy powder having 3 pairs of small black legs. It has piercing and sucking type of mouth parts. The female moth survives up to a month.

The life cycle is completed in about 9.5 to 11 months and only one generation is found in a year.

Control

1. Raking of the soil around the base of the infested tree

so that egg masses get exposed to the sun and may be killed by heat in the month of May and June.

2. Application of sticky band around the tree trunk so as to check the nymphs from crawling up the trees (4 parts castor oil + 5 parts resin) 1/2 to 1 meter above the ground level during the second week of December. It remains effective for a period of two weeks after which it should be repeated.
3. 2% methyl parathion dust may be dusted in the soil around the tree trunk in 1 meter diameter and also on the trunk up to height of 1 meter.
4. To kill the nymphs which are on the tree any of the following insecticides may be sprayed in a tree after mixing in 15-20 litres of water.
   (a) Phosphamidon 85% — 6 ml.
   (b) Malathion 50 Ec — 50 ml.
   (c) Endosulfan 35 Ec — 20 ml.
   (d) Formothion 25 Ec — 30 ml.

Natural enemies : Some Chalcid wasps and *Auliis vestila* have been found parasitising the nymphs.

Mango Leaf Hopper

Local name : Aam ka Chenpa, Pharka, Lassi Thala, Tela

Scientific name : *Amritodus (Idiocerus) atkinsoni* Leth

Other species : *Idiocerus clypealis* Leth.

I. niveosparsus Leth.

Order : Hemiptera

Sub-order : Homoptera

Family : Jassidae

Host plant : Mango.

Distribution : It is found in India, Sri Lanka, Mayanmar, Malaya, Formosa and Australia etc. In India it is commonly found in Bihar, Uttar Pradesh, Punjab, Mumbai, Andhra Pradesh and Madhya Pradesh.

Nature of damage : The damage is caused both by the adult hoppers and their nymphs in way of sucking the sap from the new shoots, buds and flowers, resulting the entire inflorescence and even small fruits to dry up, nymphs cause more damage than the adults as they are devoid of wings hence can not fly, therefore, suck the sap from a particular place for long time. They also secrete honey-dew which interferes with the respiration of the leaves. The heavy infestation of the pest usually occur periodically after every three or four years that is why the pest is known as periodic insect pest. It is suspected that one of the causes of the irregular bearings in mango is the periodical attack of this pest. Its attack starts from the end of the March and continue till the end of June. The old trees are more damaged than the newly planted trees.

Life history : Three developmental stages namely egg, nymph and adult are found in its life cycle which are described as below :

Egg : The adult hoppers hibernate in winter and become active in the beginning of February. After 3 days of copulation female lays eggs singly in the tissues of leaves, buds and flowers. The eggs are cylendrical dull white in colour and measure 0.9 mm, long. A female can lay 200 eggs in her life-span. The eggs hatch in 7-10 days.

Nymph : The newly hatched nymphs are yellowish green in colour and measure 1 mm long. They suck the sap from the tender shoots, flower buds and flowers and during this period they move about on the tree but do not fly. They moult 5 times grow in size and after 3-4 weeks of feeding they become adult.

Adult : The insect is triangular in shape being broader at anterior end and narrower towards posterior. It is 5 mm long and dull brown in colour. They cluster on the stems and thicker branches. *Idiocerus clypealis* is smaller than the above and generally rests on the under surface of the leaves. The adults of both the species have two pairs of transparent wings and 3 pairs of legs. The anterior ones are smaller and poserior ones are larger.

The life cycle is completed in 31-41 days. There are two generations in Northern India. The adult hoppers crowd on the stems and leaves, mostly on the under surface and fly about in all direction on the disturbance from middle of April to June. Then-population decline in rainy and winter seasons. They prefer shady and damp places for hide. They become again active in the commencement of spring.

Control

1. The Plants should be planted at proper distance and garden should be open to maximum sunlight and air by cutting unwanted branches trees and bushes.
2. The plants may be sprayed with any one of the following insecticides. The given amount of insecticide may be mixed in 15-20 litres of water and sprayed on a tree.
   (a) Phosphamidon 85%-6 ml.
   (b) DDVP 76 EC-10 ml.
   (c) Malathion 50 EC-50 ml.
   (d) Endosulfan 35 EC-20 ml.
   (e) Formothion 25 EC-30 ml.
   (f) Carbaryl 50 WP-25 g.
   (g) Dimethoate 30 EC-25 ml.

Natural enemies : The following are the natural enemies of the pest:

Nymphal Parasite

(a) *Pipunculus sp.*

(b) *Epipyrops sp.*

Adult parasite : *Dryinus sp.*

Bark Eating Caterpillar

Scientific name : *Inderbela quadrinotata* Walk. *I. tetraonis* Moore

Order : Lepidoptera

Family : Tetragridae

Host plants : Mango, guava, citrus, pomegranate, jamun, ber, litchi and rose etc.

Distribution : The pest is widely distributed in India, Mayanmar and Malaysia. In India it is found all over the country with particular reference to the Vidarbha region of Maharashtra.

Nature of damage : The caterpillar eats the bark and bores in the stem near the junctions of branches. It is very harmful to the tree as it feeds on the bark and in this process it seriously injures the plant vessels through which nutritive plant-sap is transported within the plant's system; the result is that the tree-growth and fruit bearing capacity are adversely affected. Sometimes the infested branches can dry up and in cases of severe infestation, the whole tree may die. The dirty elongated zig-zag ribbon-like messy web consisting of bits of bark pieces, excreta, etc. covering the trunks and branches of trees may be seen. They prefer older trees than younger ones. The infestation in some areas of the country has been reported to as high as 40 per cent.

Life history : The four developmental stages namely egg, larva, pupa and adult are found which are described as below:

Egg : The egg laying takes place in May and June. The female moth lays a very large number of eggs in small groups of 15-20 eggs in each. They are laid in crevices of bark and incubation period is about 10 days.

Larva : The larva soon after hatching begins to feed on the bark and prepares the web under which it lives. They remain hidden during the day but come out at night for feeding on the bark and they do so under coverings of webbings containing frass, faecal pellets etc.

The larval period is quite long and the caterpillar continues its destructive activity from May-June to April the following year. During this period, it grows slowly being about 1.5 cm in September and about 4 cm by December when it is practically full grown. Thus, the larval period extends over several months.

Pupa : The pupation starts from April onwards. It takes place within the larval gallery. At the end of pupal period, the pupa wriggles up to the opening of the gallery where from the moth emerges, leaving the pupal skin protruding from the exit hole. Pupal period lasts 3-4 weeks.

Adult : The adult is large-sized moth with a wing-span of about 4 cm in the female and about 3 cm in the male. It is light grey to light brick-red in colour and with dark brown patches or dots. The moth's life is quite short and after finishing egg-laying within 2-3 days, it dies.

The life cycle is completed in a year, therefore, only one generation is found per year.

Control

1. Clean the web and fumigate the hole with an ordinary fumigant by putting in the hole a swab of cotton wool dipped in the liquid fumigant (petrol, kerosine, carbon-di-sulphide) and inserting the same in the whole which should thereafter be plugged with mud.

2. Drenching the tree trunks and branches with any one of the following insecticides in the month of June :

(i)	Endosulfan	: 0.05%
(ii)	Carbaryl	: 0.15%
(iii)	Telodrin	: 0.03%
(iv)	Chlorpyrifos	: 0.03%

Mango Stem Borer

Scientific name : *Batocera rufomaculata* Dejean.

Order : Coleoptera

Family : Cerambycidae

Host plants : Mango, jackfruit, fig and mulburrey etc.

Distribution : The pest is widely distributed throughout India.

Nature of damage : Damage is caused by the grubs. They tunnel into the branches and trunks of the tree and feed on the wood inside. Sometimes they also attack the plant near ground level and even on the roots. The tunnels, which are 3 cm or more in cross-section, are packed with coarse chips of wood and fibre. Though the external symptoms of attack are not always visible, the site can be located from the sap on frass that comes out of the hole. During severe infestations, the branches are made completely hollow and therefore, easily break off or dry up.

Life history : The four developmental stages, *viz.*, egg, larva, pupa and adult are found in its life cycle which are described below :

Egg : The beetles generally appear during the monsoon and deposit eggs in the bark of trees, specially such trees which are dead or unhealthy or have wounds or cuts in them. The hatching period varies from 5-8 days.

Larva : The newly hatched grubs bore into the wood and make irregular channels inside. The grubs possess strong biting mouth parts and penetrate into the stem or even the roots and feed on woody tissues. The winter is passed in the grub stage inside the burrow. When winter is over, they again begin to feed in the spring. The full grown larva is leg less, stout, yellowish-white, fleshy and measuring about 6 cm in length. Its head is dark, with well developed mandibles. The larval stage lasts for 10-12 months and sometimes even more than this.

Pupa : The full grown grubs hollow-out a cell for pupation inside the stem. The pupal period lasts for 3-4 weeks and after which the adult beetles emerge from circular exit hole in the trees.

Adult : The adult beetle is about 5 cm in length and 2 cm in breadth. It has long horn and pale-greyish in colour, with several creamy, irregular spots on its back. The head is distinct, with large prominent eyes, and the pronotum is ornamented with two crescent orange-yellow spots. The beetles are nocturnal in habit and feed to some extent on the bark of the living twings etc.

The life cycle is completed in 12 months and only one generation is found in a year. During the non-active season, the pest takes shelter under the bark of the mango tree or cracks and crevices in the adult stage. Adult beetles generally appear in May and June.

Control

1. Prune and destroy all the affected branches during winter.
2. In case of stems, insert the soacked cotton in fumigants like carbon-di-sulphide etc. in the holes.
3. Drench the tree trunks and branches with 0.05% endosulfan or 0.03% telodrin emulsion in the month of June.

Anar Butterfly

Scientific name : *Virachola isocrates* Fabr.

Order : Lepidoptera

Family : Lycaenidae

Host plants : Pomegranate, guava, loquat, orange and tamarind.

Distribution : Throughout India.

Nature of damage : The anar butterfly is an important pest of pomegranate and causes heavy damage to its cultivation and causes heavy damage to its cultivation. The larva is a damaging stage, which bores into the fruit and feed on its contents. It creates a lot of mess and offensive smelling matter oozes out from the enterance hole. The wound caused by this pest gets infected with a number of bacteria and fungi thereby making the fruits rot. The holes made in the fruits by larvae are usually visible. During severe infestation the loss caused by the pest is estimated to the tune of 90%.

Life history : There are four developmental stages are found in its life cycle which have been described below :

Egg : The female butterfly lays whitish, shiny egg singly on the lower ends of the flowers, on the fruit surface and sometimes even on the leaves near the fruits. The hatching period of eggs varies from 7-10 days.

Larva : The young caterpillar instead of feeding on the leaves bores into the fruit and feed inside. It first feeds on the pulp and then to the seeds. When the larva is full grown, it comes out from the hole bores through the hard shell of the fruit and spins of web by which it at taches the fruit secured to the stem. This is done to prevent the fruit from falling down. The caterpillars are short, dark brown in colour with hairs and whitish patches on the body. The larval period lasts for 18-47 days. A single fruit may have about half a dozen caterpillars.

Pupa : The full grown caterpillars pupate either in the damaged fruits or on the stems holding them. The pupa is dark brown in colour. The pupal period varies from 7-34 days depending upon environmental conditions.

Adult : The butterfly is about 2.5 cm across the wings and glossy violet in colour in case of male and violet brown in case of female. The hind wings possess a tail like prolongation. The eyes have a rim of white scales.

The life cycle is completed in 33-93 days and breeding of the pest continues throughout the year with varying speeds depending on environmental conditions. There are 3-4 generations in a year.

Control

1. Bag the young fruits with stiff paper or polythene or coarse paper.
2. Spraying the trees, repeatedly with 0.07% endosulfan or 0.15%, carbaryl or 0.03% phosphamidon.
3. Destroy the infested fruits to minimise the damage.

Ber Fruit Fly

Scientific name : *Carpomyia visuviana* Costa

Order : Diptera

Family : Trypetidae

Host plant : Ber.

Distribution : The pest is distributed in all parts of India, West Pakistan and the Middle East extending upto Southern Europe.

Nature of damage : The damage is caused by the maggots of the fly which feed on the pulp inside the fruits. The fruits attacked by the pest become unfit for human consumption.

During severe infestation usually in the month of February, it is difficult to get even a single sound and healthy fruit from hundreds of trees. During the peak period of attack the number of maggots is more than one inside the fruit and the whole pulp is consumed and decay set in by the maggots with the result that when damage has been completed, nothing but a semi-solid decayed dark brownish mass which smells offensively, is found within the fruit.

Life history : The four developmental stages namely egg, larva, pupa and adult are found in its life cycle, which have been described below :

Egg : The eggs are laid by the female fly singly or in clusters of 4-5 under the skin of the fruit and from the punctures thus made oozes out a little fluid which often dries up in the form of globule or dot of resinous material on the fruit. The hatching period varies from 4-5 days.

Larva : The young maggots after hatching from the eggs bore into the pulp and feed. The full grown maggot is amphineustic, creamy white pointed anteriorly and ends in a broad truncate last segment. The dorsal vessel is prominent. The full grown larva measures '6 mm in length and 1.5 mm in breadth. The maggots moult two times in its life time and larval period lasts for 9-12 days in March and 22 days in November-December.

Pupa : The maggots before pupation greatly contracts and thickens and pupates in the 1st larval skin which forms the puparium. Pupation generally takes place in the soil at a depth of 4-6 cm. usually the maggots leave the fruit for pupation out side the fruit but sometimes the drying state of the fruit may prompt the maggots to pupate within the fruit also. The pupal period lasts from 11-31 days during March-April and 45 to 87 days in winter. The well developed pupa measures 4.5 mm in length and 1.80 mm in breadth and light brown in colour.

Adult : The head of the freshly emerged fly is light almond brown. The wings are hyaline and transparent. The fly is about the size of a housefly, slightly more slender, with a pale, brown body, with an oval whorl of brownish black spots behind the head.

The pest is active in autumn and spring and the summer and winter are passed in the pupal stage. The total life cycle is completed in 25 to 117 days and there are 3 generations in a year.

Control

1. The attacked fruits should be collected and burnt or fed to the cattle.
2. Spray the trees with 0.07% endosulfan or 0.15% carbaryl before the fruits mature at intervals of 15 days.
3. Wild ber bushes should not be allowed to grow near the ber plantations.

Citrus Leaf Miner

Scientific name : *Phyllocnistris citrella* Staint.

Order : Lepidoptera

Family : Gracillariidae

Host plants : Citrus, pomelo, willow, cinnamon and *Loranthus* spp.

Distribution : It is found in all parts of India and occurs in many countries of South East Asia and Australia It is known to be a serious pest of citrus nurseries in Tamil Nadu, Madhya Pradesh, Assam, Uttar Pradesh and Punjab.

Nature of damage : The caterpillar causes damage by making zig-zag silvery mines in the young leaves. Due to the extensive mining by the pest, the leaf suffers badly, gets deformed and irregularly curled up in shape, unhealthy in look and defective in its function and finally it dries and falls off. On the older leaves, brownish patches are formed which serve as foci of infection for citrus canker.

The succulent leaves with thin epidermis are more suitable for the penetrating and mining activities of the tiny larvae. The larvae feed more on plant sap than on the tissues. As a result, the leaves turn pale, curl up badly and dry off. In case of the nursery plants the entire stock may be ruined. In case of larger trees, photosynthesis is adversely affected due to which an appreciable reduction occurs in yield.

Life history : The four developmental stages namely egg, larva, pupa and adult are found in its life cycle, which are described below :

Egg : The eggs are usually laid singly on the underside of the leaves and tender shoots. They are mostly found 2 to 3 in each leaf. The eggs are minute, flattened, transparent and mostly laid near the mid rib. The hatching period varies from 2-10 days depending upon the environmental conditions.

Larva : The newly hatched larva is legless which enters the leaf tissue and begins to feed inside it and mines the leaf lamina. They form galleries within which they remain confined for the rest of their immature life. The full grown larva measures 5.1 mm in length and is pale yellow or pale green with light brown mandibles which are well developed. The larval period lasts 5-30 days.

Pupa : The full grown larva comes-out and pupates near the margin of the leaf which folds upto provide a sort of cover over the pupa. In the form of its body, the pupa is slightly brownish. It is partly exposed through the gallery wall and has a spine on its head with the help of which it pierces through the wall when it emerges as a moth. The pupal period varies from 5-25 days and the moths are commonly seen resting on the trunks of the trees near the ground.

Adult : The adult is a tiny greyish moth with a wing span of 8 to 10 mm. The forewings are white with two narrow grey stripes and the hind wing has a pale grey fringe.

The life cycle is completed in 14-68 days depending on the environment. The cycle of generations continues practically the whole year except that the development gets prolonged in the colder months. The moth activity may cease if it is very cold during December and January.

Control

1. Collecting and burning the mined leaves found useful.
2. The orchard should be kept free of wild citrus hedges, which serve as breeding places of the pest.
3. Spraying the trees with anyone of the following insecticides at 15 days intervals :

   (i) Methyl demeton : 0.03%

   (ii) Phosphamidon : 0.03%

   (iii) Dimethoate : 0.03%

   (iv) Monocrotophos : 0.04%

Citrus Psylla

Scientific name : *Diaphorina citri* Kuwayanae.

Order : Hemiptera

Family : Psyllidae

Host plants : All species of citrus and plants of the family Rutaceae.

Distribution : It is found in several parts of India and in the neighbouring countries, *viz.*, China, Formosa, Japan, Mayanmar, Sri Lanka, the East Indies and New Guinea. In India it is most destructive in Punjab, Himachal Pradesh and Maharashtra.

Nature of damage : Damage is caused by the nymphs and adults. They suck the sap of the fresh tender parts of the host. The vitality of the plants deteriorates and the young leaves and twigs stop growing further. The leaf and flower buds may wilt and die. The small fruits formed in the spring fall off prematurely. The honey-dew excreted by the nymphs acts as a focus for the growth of the moulds on the leaves which adversely affects the photosynthesis. It is also thought that the insect produces a toxic substance in the plants with the result the fruits remain undersized, poor in juice and are insipid. The pest is also responsible for spreading the greening virus. If the pest is not controlled in time, the entire orchard may be lost.

Life history : There are three developmental stages namely egg, nymph and adult found in its life cycle, which are described below :

Egg : The adults start breeding in February-March and lay on an average 500 almond-shaped, orange and stalked eggs on tender leaves and shoots of citrus trees. The lower end of egg stalk is embedded in the plant tissue. They are laid either singly or group of two or three which are arranged in straight lines. There may be as many as 50 eggs in a place. The hatching period varies 4-6 days in summer and 10-20 days in winter.

Nymph : The newly hatched nymphs are light yellow in colour and have a tendency to stick close to the egg-shell. The nymphs are flat, louse-like and are seen congregated in large numbers on the young leaves and the buds, but in the later stage they may migrate to the older leaves. There are five nymphal stages and the development is completed in 12-25 days which may extend 34-36 days in December-January. When the nymphs are full grown, they migrate to the lower surface of the leaves where they change into adults.

Adult : The adult is small, 3 mm in length, active in habits, and rests on the leaf surface with closed wings. The insect is brown with its head lighter brown and pointed. The wings arc membranous, semi-transparent, with a brown band in the apical half of the fore wings. The hind wings are shorter and thinner than the fore wings. The females live longer than the males.

The adults copulate 4-8 days after emergence and the females start egg laying immediately. The life cycle is completed in 20-64 days depending upon the environmental conditions. There are 8-9 overlapping generations in a year. During severe winter only adults survive, similarly in hot period of May-June, the adults are predominant, although other stages may also be found.

Control : The pest may be controlled by spraying the trees with any one of the following insecticides :

(i) Phosphamidon : 0.03%

(ii) Methyl demeton : 0.05%

(iii) Dimethoate : 0.03%

(iv) Malathion : 0.05%

(v) Monocrotophos : 0.04%

Lemon Butterfly

Common name : Neebu ki surhi, Neebu ki titli.

Scientific name : *Papilio demoleus* Linn.

Order : Lepidoptera

Family : Papilionidae

Host plant : Citrus.

Distribution : It is widely distributed in the Indian subcontinent and is found right from Arabia in the West, to Formosa in the East.

Nature of damage : The larva of the butterfly is damaging stage which feeds on the leaves of citrus plants of all kinds.

The caterpillar cuts the leaves irregularly in the nursery. The leaves of young plants are completely eaten up. Thus the young plants suffer more by the attack of this pest. In severe infestation the larvae may eat even tender shoots. The pest is generally active from March to November and severe outbreaks generally occur during August-September.

Life history : The following four developmental stages are found in its life history :

Egg : The female butterfly lays yellowish-white eggs on young leaves and tender shoots. The eggs are smooth round and are generally laid scattered and singly. Sometimes they may be found in group of 2 to 5. A female can lay 75-120 eggs in her life span. The hatching period of eggs varied form 3-4 days in summer and 5-8 days in winter.

Larva : The newly hatched larva is about 3 mm long and initially feeds on the egg-shell and this habit persists through out its life. Larva eats exuvae after each moult. The young caterpillars are brown with white irregular markings in the body surface and they look like as if they are not caterpillars but some irregular masses of bird excreta. Obviously, this is an adaptation to escape the notice of their enemies like birds who would not like to pick-up their own droppings. The larvae become full grown in 8-16 days during summer and in about 28 days during winter. The full grown larva is yellowish green in colour and measuring 40 mm in length and 6.5 mm in breadth. It possesses a horn like structure on dorsal side of last abdomninal segment.

Pupa : The pupation takes place on the leaves or stem with a silken thread secreted by it. The pupa is husky in colour and passed winter in this stage. The pupal period varied 7 days in summer and 56 to 98 days during winter.

Adult : The butterfly is large and conspicuous with black and lemon yellow marks on its wings and with two blue eye like marks on the hind wings in the anal region. It is about 28 mm long and has black clavate type of antennae.

The butterfly after one day of emergence copulates and starts egg laying on 2nd or 3rd day.

The life cycle is completed in 20-117 days depending on environmental conditions and many generations are found in a year.

Control

1. During severe infestation the caterpillars may be picked and destroyed.
2. Any one of the following insecticides may be sprayed on a tree after mixing in 15 lit. of water.
   - *(i)* Endosulfan 35 EC — 25-30 ml.
   - *(ii)* Phosphamidon 85% — 5-6 ml.
   - *(iii)* Formothion 25 EC — 30 ml.

Natural enemies : *Ericia pimphalidophaga* is a larval parasite.

Singhara Beetle

Scientific name : *Galerucella brimanica* (Jacoby)

Order : Coleoptera

Family : Chrysomelidae

Food plant : Singhara.

Distribution : It is widely distributed in Pakistan, Sri Lanka, Mayanmar and India.

Nature of damage : Damage is caused by both grubs and beetles which feed on the leaves of waternut floating on the surface of the water. The attacked leaves are riddled with holes and rendered useless, as a result of which the growth of the plant is checked and the formation of flowers and the production of fruits greatly reduced. Sometimes the integument of *singhara* fruits at the surface of the water is also damaged which results in the rotting of the fruits. A badly damaged crop turns black instead of looking green maximum damage is caused by the pest during August and September.

Life history : The four developmental stages namely egg, larva, pupa and adult are found in its life cycle which have been described as below :

Egg : The beetle lays eggs on the upper surface of the leaves in clusters of 10-20. The eggs are rounded and reddish-brown in colour. One female may lay as many as 300 eggs during its life span. The eggs are securely glued on to the leaves. The hatching period of eggs varies from 6-9 days.

Larva : The newly hatched grubs start feeding on the upper epidermis of the leaves. The young grub is light brown in colour and undergoes in moultings. The full grown larva is dark brownish-grey in colour and about 6-7 mm long. The larval stage lasts for 10-22 days.

Pupa : Before, pupation, the grub becomes sluggish and enters the pre-pupal stage lasting for 1-2 days. The pupation takes place on the leaf surface and like the eggs the pupa are also firmly flued to the leaves. The pupa is oval shaped and orange red in colour with dark brown patches. The period lasts 3-7 days after which the adult beetle emerges out.

Adult : The adult beetle is blackish-brown in colour, 6 mm in length and 3 mm in breadth. It possesses a large hump in the middle of the body and eyes are black. The adult beetles live for a month or so, although some of them may live for three months.

The pest is active generally throughout the year but in places where water is receded, it aestivates in cracks and crevices in the sides of ponds. These beetles again become active in July. The total life cycle is completed in 21-41 days and there are generally 6-7 generations in a year.

Control

1. Collect egg masses and beetles from the leaves and destroy them in the early stage of the crop.
2. Dusting the crop with any one of the following insecticides per hectare :
   - *(i)* Malathion 5% dust : 25 kg.
   - *(ii)* Carbaryl 10% dust : 25 kg.
   - *(iii)* Carbaryl 5% dust : 30-35 kg.

Fruit Sucking Moths

There are about 20 species of fruit sucking moths out of which following are more important

Ophideres conjuncta (fullonia) Cramer

O. matema Cramer

Achoea Janata Linn.

Calpe emarginata Fabr.

Common name : Neebu ke phalon ka patinga

Scientific name : *Ophideres* spp.

Order : Lepidoptera

Family : Noctuidae

Host plants : Citrus, grape-fruit, pomegranate, malta and orange. The caterpillars also feed on the wild plants.

Nature of damage : Generally moths and butterflies damage the plant or plant products only during their larval stages but fruit sucking moths as their common name indicates damage fruits in their adult stage. The moths damage the outer skin and then suck the juice of fruits. Usually ripening fruits are attacked but unripe fruits are not always immune from such attacks. Severely attacked fruits rot and fall to the ground. The damage also exposes the fruits to be attacked by fungi and bacteria etc. The damage caused by them is estimated to the tune of 20-40 per cent.

Life history : The four developmental stages *viz.*, egg, larva pupa and adult are found in its life cycle which are described as under :

Egg : The female lays eggs singly on the underside of the leaves of creepers namely gurch *(Tinospora codolia)* and vasnable *(Cocculus villosus)*. The eggs are round or oval measuring 1 mm in diameter and shining pale-green in colour. A female lays 200-300 eggs in her life span. The eggs hatch in 8-10 days.

Larva : The larvae start feeding on the host plants (wild creeper and plants) within 24 hours of hatching. The caterpillars undergo 5 moults and become full grown in 12-21 days. The full grown caterpillar is stout, hump backed (semi looper) and measures SO to 60 mm in length. The larva is velvety-blue in colour with dorsal and lateral sides covered with bright-red and yellow spots.

Pupa : It pupates in a leaf-fold under a very thin whitish-pale patchment made up of silken threads. The pupa is stout and reddish-brown in colour. Pupal period lasts from 8 to 15 days.

Adult : The adult moth is 2.5 cm long with a wing span of 9 cm. The fore wings are brownish green while hind wings are orange or yellow in colour with black spots in case of *O. conjuncta*.

The moths of *O. matema* are light orange to brown in colour. Fore wings are dark brown while hind wings are orange-red in colour with two black spots.

The fore wings of *O. ancilla* are yellowish-green with dark markings and white spots.

The life cycle is completed in 31-50 days and generally 3

generations are found a year. The moths are nocturnal in habit and remain active during the rainy season, mainly from July to September.

Control

1. The wild creepers, plants which are the food plants of larvae should not be allowed to grow near the orchards.
2. The moths may be attracted in light traps and destroyed.
3. Smoking in the orchards escape the odour of ripening fruits which attract the moths.
4. The fruits may be bagged with plastic or cloth bags.
5. Bait the moths with lead arsenate and gur solution mixed in the proportion of 1 : 160 and hang on the branches in small pots. Few drops of vinegar may be mixed to attract the moth.

Castor Semilooper

Scientific name : *Achoea janata* Linn.

Order : Lepidoptera

Family : Noctuidae.

Host plants : Adult moths suck the sap of ripening fruits while larva feeds on mainly castor leaves.

Distribution : It is distributed through out the Indian continent.

Nature of Damage : The moths also damage castor plants and guava fruits besides citrus fruits. The moths pierce the fruits and suck the juice from them. One fruit may have as many as 20 to 25 punctures in it as a result of which it rots and falls to the ground. The infested fruits are also attacked by fungi and bacteria etc. which initiate early roting, the larva also causes heavy loss to castor plant during severe infestation the entire leaves are eaten up leaving veins and stalks only.

Life history : There are four (developmental stages namely egg, larva, pupa and adult in its life cycle described below :

Egg : The female moth lays eggs scattered on the under surface of leaves. The egg is comparatively large (0.9 mm long) and has ridges and furrows on the surface radiating from a circular depression at the apex. It is blue-green in colour and difficult to spot out on the leaf. A female lays about 400-500 eggs in her life time. Hatching-period lasts from 2-5 days.

Larva : The larva emerges by biting a hole to the egg-shell which forms the first meal of the newly hatched larva before it feeds on the leaf tissue. The first stage larva is about 3.5 mm in length. It moults 4 to 5 times and becomes full grown measuring 60-70 mm in length. The locomotory process involving the formation of a semi-loop of the larval body. On the dorsal side of the larval body white and red stripes are found. The larval period lasts from 5 to 20 days.

Pupa : Pupation takes place both in the soil and under fallen leaves and other rubbish at the edge of the field. The larva prepares a loose cocoon of coarse silk to which particles of soil readily adhere. Some times pupation also takes place within the folded leaves on the plant itself. The pupal stage lasts from 10 to 15 days.

Adult : The moth is pale-reddish brown in colour with wing span of 60-70 mm. Both the fore and hind wings have broad zig-zag decorative markings and large pale dark brown patches. They are nocturnal in habit and attracted in light.

The winter is passed in pupal stage and moths emerge out in late February or in early March. The female starts egg laying after 3-4 days of copulation. The life cycle is completed in 30-44 days, there are 5 to 6 generations in a year.

Control : The moths may be controlled in a similar way as mentioned in the case of *Ophideres* spp. The larvae attacking castor crop may be controlled as follows :

1. Dusting the crop with 5 or 10% BHC dust @ 25 or 30 kg/hect. found effective.
2. The spraying of the crop with any one of the following insecticides has also been found useful :
   (a) Endosulfan 35 EC - 1.25 lit/hect.
   (b) Phosphamidon 85% - 225 ml per hect.
   (c) Methyl parathion 50 EC - 1 lit/hect.

Potato Tuber Moth

Common name : Alu-ki-surhi

Scientific name : *Pthorimoea (Gnorimoschema)operculella* Zeller.

Order : Lepidoptera

Family : Gelechiidae

Host plants : Potato, tomato, brinjal and tobacco etc.

Distribution : This is found in India, China, Australia, New Zealand and Hawaii. In India it is distributed every where except the hilly tracts of the country.

Nature of damage : The damage is caused by larvae which bore in to the tuber and feed on the pulp. Damage is done in the field as well as in godowns but it is a serious pest of stored potato. In the field it causes damage from October to February-March. Initially the caterpillars mine the leaf bore in to branches and stem and feed inside but during severe infestation they may eat leaves also. The exposed tubers in the field are also subjected to their attack. With these infested potatoes they reach to godown and multiply. The damaged tubers are unfit for seed as well as for human consumption. In addition, the attacked tubers are exposed to bacterial infection which leads to rotting. The infested tubers may be differentiated by the presence of black excreta near the eye bud. Storing of potatoes in cold storage reduces the tuber moth attack almost completely.

Life history : The four development stages, *viz.*, egg, larva, pupa and adult are found in its life history which are described as under :

Egg : The female moth lays eggs in or around eyes or cracks of skin of the tubers in godowns. In the field eggs are laid on the lower surface of the leaves. The eggs are pearly-white in colour when freshly laid, turning brownish or blakish lateron and measuring 0.6 mm in length and 0.4 mm in breadth. A female lays 100 to 150 eggs in her life span. The eggs each in 3 to 4 days.

Larva : The newly hatched larva is about 1.5 mm long white in colour with black head.

It bores in to the tuber and enters in it through a minute hole. The caterpillar makes tunnels in all directions in the pulp of the tuber and becomes full grown in 10-20 days. It moults 5 times and full grown caterpillar is pinkish in colour and measures about 22 mm in length. A single potato may have several caterpillars. The full grown larva comes out of the tuber for pupation.

Pupa : The larva pupates among the potatoes or crevices of their containers or on fallen leaves in the field. It forms a dirty cocoon of 6.5 mm long for pupation. The pupa is brown and measures 6.5 mm in length and 2.5 mm in breadth. Pupal stage lasts from 7 to 10 days.

Adult : The adult moth is greyish-brown in colour measures 6 mm long with a wing span of 12 mm. The fore wings are greyish-brown with yellow longitudinal bands and hind wings have a fringe of hair on the posterior margin. The adults are nocturnal in habit and attracted in the light.

The mating of moths takes place soon after their emergence and female starts egg laying after 1 or 2 days of copulation. The life cycle is completed in 21-36 days and 6 to 8 generations are found in a year.

Control

1. Affected tubers should be picked and rejected before storage.
2. The potatoes are stored in sand (2.5 mm thick) so that the female could not lay eggs on them.
3. In the field when infestation starts, crop may be sprayed with endosulfan 35 EC @ 1 lit/hect. in 750 lit of water.
4. If the potatoes are stored for seed purpose the malathion dust may be mixed @ 1.5 kg/m tonnes of tubers.
5. Fumigation with aluminium phosphide @ 7 tablets/28 cubic meter space is effective in killing the caterpillars.

Natural enemies : The following are the effective larval parasites :

(i) *Bracon hebitor*

(ii) *Serratia marcescens*

(iii) *Microbracon gelechiae*

Rice Weevil

Common name : Ghun, Killa, Ghanera, Sund Wali Surasari.

Scientific name

(i) *Sitophilus oryzae* Linn.

(ii) *Sitophilus granarius* Linn.

Order : Coleoptera

Family : Curculionidae

Food : Rice, paddy, wheat, barley, maize, jowar etc.

Distribution : It is cosmopolitan in distribution but is generally much more injurious in warm humid countries. India is considered to be its original place.

Nature of damage : This is the most destructive insect pest of stored cereals in the world. It is called the rice-weevil only because it was found to be infesting rice when it was described for the first time. Both, adults and the larvae (grubs) damage the grain on which they feed voraciously so mesh so the grain is rendered unfit for human consumption as well as for seed purposes. The grub is more injurious than adult it makes a tiny hole in to the grain, enter inside and feed on the content of the grain leaving only shell.

Usually it is active throughout the year but serious damage is caused from July to November. Moisture is important factor for its development. Under favourable conditions, it is found in great number and due to respiration heat spots are found in the grains. In cases of heavy infestation the grain becomes a mass of broken vegetable matter. Some times black fungus also grows.

Life history : The four developmental stages namely egg, larva, pupa and adult are found in its life cycle which have been described as below :

Egg : Eggs are deposited by the female after 5 days of adult formation. The mother-weevil makes a small cavity in the grain by means of its powerful jaws and then lays an egg in the cavity and cover it with a gelatinous fluid. Usually one egg is laid in a grain but some times 2 or 3 eggs may be deposited. The eggs are white in colour, oval in out line and measures about 0.7 mm long and 0.3 mm broad. A single female can lay 250-300 eggs in her life span. The eggs hatch in about two days during summer and 6-9 days during winter.

Larva : The tiny grub immediately after hatching bore into the grain kernel. The grub is white in colour with yellowish brown head legless and provided with biting type mouth parts. It remains inside the grain feeds on starchy content and emerges on reaching the adult weevil stage. It moults 3 times inside the grain and becomes full grown in 25-35 days. The full grown grub is yellowish-white in colour and about 0.3 cm in length.

Pupa : The grub makes a pupal shell inside the grain and pupates therein after passing a day or two as pre-pupal period. The pupa is at first dirtywhite in colour but later on it changes to dark brown and measures 2.5 mm to 3.0 mm in length. The pupal stage lasts for 4-7 days during summer and 14-20 days during winter.

Adult : It is a reddish-brown weevil which turns into black brown, about 3.5 mm in length and its head has a slender pointed forward projection (snout) with a pair of stout mandibular jaws at its extremity. The snout is about 1 mm long possess clavate antennae at its base. The adult of *S. oryzae* are light brown in colour having 2 pairs of spots on each elytra while these are absent in S. *granarius.* The S. *granarius* can not fly as its wings are fused while *S. oryzae* is strong flier.

The life cycle is completed in 36-69 days and 6-8 generations are found in a year. The adult weevil may live 21-28 days in summer and 3 months in winter. The weevil hibernates during severe winter in cracks and crevices.

Its life history is mainly dependent on humidity and temperature. At 70% humidity and 17°C temperature it is completed in 220 days, as the humidity increases it takes much less time to complete the life cycle. During rainy season, when conditions are favourable a pair of weevil may produce 10,00,000 insects in a season.

Control : Control will be given in last for all the storage pests.

Natural enemies : The following are the important natural enemies :

Egg parasite : *Acarapsis doctamite*

Grub parasite : Larrophagous distingnendus forst

Aplastomorpha calanrae how

Pendiculeides ventricosis new port-mite.

Khapra Beetle

Common name : Khapia, Pat, Wantari.

Scientific name : *Trogoderma granarium* Everts

Order : Coleoptera

Family : Dermestidae

Food : Wheat, Rice, Jowar, Bajra and Maize etc.

Distribution : It is mainly distributed in India, Africa, Mayanmar and China and also found in countries where temperature ranges from 32-44°C.

Nature of damage : Unlike other beetles the adult in this case is harmless and the damage is caused only by the larval stage. The grub first attacks at embryo point but later when infestation is severe other parts of the grain are also badly damaged. Usually the infestation occurs in superficial layers of the grains, as the pest do not penetrate beyond some depth. This is due to the fact that in its egg pupal and adult stages the beetle is highly susceptible to reduction in oxygen tension. Maximum damage by the pest is caused during July to October. Since it mainly feeds on embryo, germination power of infested grain is adversely affected.

Life history : There are four developmental stages, *viz.*, egg, larva, pupa and adult in its life cycle, which are described as under :

Egg : The female starts egg laying after 5-6 days of copulation. The eggs are generally laid singly or rarely 2-5 together among the grain. An individual egg is round small, semi-transparent white in colour and measures 1/2 mm long. The eggs are laid sometimes as many as 26 per day and a single female can lay about 125-150 eggs in her life time. The eggs hatch in 6-16 days depending upon the environmental conditions.

Larva : The newly hatched grub is yellowish white in colour but later turn brownish with transverse yellowish-brown bands and bundles of reddish-brown hairs over and along the body. The long hairs are also found at the hind end of the body. The male grub takes 19-28 days in development while female grub 20-37 days. Under adverse conditions, they may prolong their life to as long as 4 years, living for the greater part of it even without food. Usually they moult 4 times but under adverse conditions they moult 10-12 times. It passes winter in its grub stage and remains hidden in cracks and crevices. Full grown grub is about 5 mm long.

Pupa : Pupation takes place in the last larval skin among the grains or in cocoons. The pupa is dark brown in colour and having reddish-brown hairs on it. The pupal stage lasts for 4-27 days depending upon environmental conditions.

Adult : The female beetle is a small rather oval, pale red-brown or black in colour and measures 2.5 mm in length and 2.00 mm in breadth. It has distinct markings on the wing covers. The male is almost half the size of the female. The antennae are small and capitate type.

The adults are ready to start the next generation very soon after emergence. The life cycle is completed under favourable conditions in 32-84 days. There are generally 20-22 generations in a year. The adult beetle may survive 25-35 days.

Natural enemies

(i) Reduvid bugs predator as grub stage

(ii) *Grub parasites : Anisopteromalus calandrae,* Harward, *Dinarmus laticaps, Ashmead-Synopeas* sp.

Lesser Grain Borer

Common name : Lal Surhi, thuthun heen surhi

Scientific name : *Rhizopertha dominica* Fabr.

Order : Coleoptera

Family : Bostrychidae

Food : Wheat, rice, maize, jowar, barley, gram, flour and dried fruits etc.

Distribution : It is cosmopolitan in nature and mostly found in India, Pakistan, America, Argentina and Australia.

Nature of damage : The damage is caused by both beetle and grub but beetle is more harmful than the grub. In some places it is considered to be next to rice weevil but other places it is reported to be causing more damage than even rice weevil. The grubs eat out the starchy contents of the grains leaving the outer husk only while beetles destroy whole grains which are reduced to frass and waste flour. The first stage larva is straight and can bore into sound grain but in the later stages of growth it is curved, therefore, it becomes difficult to them to penetrate the grain. The grubs which are unable to penetrate the grain feed on the waste flour left by the adults. Some times adult beetles attack the grain in the field it self. The serious damage is caused during May to August

Life history : The four developmental stages, *viz.*, egg, larva, pupa and adult are found in its life cycle which are described as below :

Egg : The beetle starts egg laying often 4-5 days of becoming adult. The eggs are laid from April to November. The female generally lay eggs on the grain near the embryo-end, which is comparatively soft and from where the newly hatched grub easily enter the grain. The eggs are pear shaped and are laid either singly or some times in clusters. They are initially white in colour which later turns into pinkish. A female can lay 300-500 eggs in her life time. The hatching period varies 5-6 days during summer and 15-21 days during winter.

Larva : The newly hatched grub is quite active and white in colour, soon enters into grain generally near the comparatively soft embryo end and passes the rest of its life therein. Some of them remain feeding on the powdered starchy material without entering the grain. The larva undergoes 4-5 moults. The full grown larva is dirty white in colour with a light brown head and curved abdomen, some what 'C shaped. It measures 2-3 mm in length and covered with tiny hairs. The grub period lasts for 40-50 days.

Pupa : Pupation takes place either inside the grain or out side the grain depending upon where grub lives. The pupa is dirty white in colour and irregular in shape. The pupal period is of 7-8 days.

Adult : It is a small polished, dark brown or black beetle measuring 3-4 mm in length and a slender cylindrical body with its head turned down under the thorax. The antennae are 10 segmented and terminate in a prominent club.

The life cycle is completed in 56-84 days and 5-7 generations are found in a year. The insect hibernates both as a grub and adult during December to February. They resume activity in March and adults starts to lay eggs in April after 4-5 days of copulation.

Natural Enemies

(i) *Lariophagus distinguendus Frost-grub* parasite

(ii) *Pediculoides* sp.—A mite predating on eggs and grubs.

Red Rust Flour Beetle

Common name : Sursali, Lalsuri.

Scientific name : *Tribolium castaneum* Herbst.

Order : Coleoptera

Family : Tenebrionidae

Food : Stored grains and stored products of various kinds such as flour, maida, suji, cotton, ground nuts, beans and dry fruits etc.

Distribution : It is found all over the world and in India the pest is mainly distributed in Uttar Pradesh, Bihar, Madhya Pradesh, Punjab, Haryana and Rajasthan.

Nature of damage : The insect does not cause damage to whole grain but mainly feeds on broken or attacked grains of other insects. It is generally found in association with Khapra beetle, rice weevil and lesser grain borer and does not cause much damage as done by these pests. The insect is specially harmful to flour, maida and suji. In cases of severe infestation flour and flour products turn greyish yellow and become mouldy which produces pungent smell and become unfit for human consumption. Both grub and adult are responsible for causing damage. The maximum damage is caused during rainy season.

Life history : The four developmental stages, *viz.*, egg, larva, pupa and adult are found in its life cycle which are described as below :

Egg : The female starts egg laying after 2-3 days of copulation. The eggs are generally laid singly in flour or among grains and being sticky, covered with flour or dust and are difficult to detect. A single female can lay 400-550 eggs in her life time which are cylindrical and white in colour. The eggs hatch in 4-12 days depending upon environmental conditions.

Larva : The young grub is small, slender and cylindrical in appearance with its body segments having a number of fine hairs and the terminal segment being in addition provided with a pair of spine like appendages. At the time of hatching the grub is 1 mm long which grows up to 5 mm when fully developed. It moults 6-7 times and becomes full grown in 27 to 90 days depending upon the food available and prevailing temperature. At 30°C temperature it becomes full grown in 22 to 25 days.

Pupa : Usually, pupation takes place on the surface of the food. Initially the pupa is white in colour which gradually turns into yellowish and covered with hairs on the dorsal surface. Pupal stage lasts for 5-9 days.

Adult : It is small reddish-brown beetle measuring 3.5 mm in length and 1.2 mm in breadth. The head, thorax and abdomen are distinct. The antennae are well developed and last three segments form an enlarged terminal end.

The life cycle is completed in 38-114 days and many generations are found in a year. The beetle hibernates during winter and resumes activity in spring when they copulate after 1 or 2 days of their emergence.

Control : Like other storage pests.

Pulse-beetle

Common name : Dhora, Chiraiya, Dhorna, Dhanur.

Scientific name : Following 3 species are found in India :

(i) *Callosobruchus chinensis* (Linn.)

(ii) C. *maculatus* (Fab.)

(iii) C. *analis* (Fab.)

Order : Coleoptera

Family : Bruchidae

Food : Arhar, mung, urd, pea, gram, lentil, cowpea, soyabean, moth and beans etc.

Distribution : It is world-wide in distribution and has been found in all parts of India. It was first described from China in 1758 due to which the name chinensis was given to the species.

Nature of damage : Generally damage starts in the leguminous pods in the field from where they are carried to storage godowns. However, they do really serious and spectacular damage under storage conditions. Both grubs and adult beetles are responsible for causing the damage. The grubs cause damage by eating out the entire content of the grain, leaving only the shell behind. The damaged grains are unfit for human consumption as well as for sowing purposes. The maximum damage is caused during July to September. The losses caused by this pest to the pulses have been estimated to the tune of 40 to 50 per cent in storage.

Life history : Four developmental stages, *viz.*, egg, larva, pupa and adult are found in its life cycle which are described as below :

In the field, the eggs are laid on the developing pods either on the outside or inside the pods. The tiny grubs after hatching bore the grains and develop inside. The entry hole is generally too small and gets healed up soon. Under storage conditions, the eggs are glued on the grains, singly but many eggs may be seen on a single grain. The eggs, when freshly laid, are translucent, smooth and shining, which later become yellowish-white and very prominent against the contrasting back ground of the grain surface. A single female lays about 60-90 eggs in her life time which hatch in 4 to 5 days.

Larva : The young grubs bore into the pod or grain and develop inside where they grown up. The full grown grub is about 5 mm long cylindrical in shape, fleshy, wrinkled and white in colour except head which is brown, in colour. It has three pairs of legs and each leg possesses a claw at the terminal end. The grub moults 4 times and becomes full grown within 2-3 weeks.

Pupa : When grub is full grown it cuts a circular disc in the seed coat which just remains in position and it pupates inside the grain. The pupa is brown in colour and pupal period lasts for 4 days during favourable conditions, which extends to 4 weeks during winter. After which the beetle emerges by further cutting out the circular disc into an emergence hole.

Adult : The adult is a small chocolate coloured beetle measuring 3.2 mm in length. The head is small with blunt snout, the femur of the hind leg specially thickened, truncate elytra which do not cover the posterior portion of the abdomen. The antennae are serrate type in case of female and pectinate in case of male.

The pest hibernates during winter as larva and adults begin to appear towards the end of March. The female starts egg laying after 3-4 days of copulation. The life cycle is completed in 25-58 days and 7 to 8 overlapping generations are found in a year. The adults are short lived and may not survive for more than two weeks.

Food : Wheat, maize, rice, barley, jowar etc.

Distribution : The pest is also known as Angoumois grain moth because it was first reported in 1736 from the Angoumois province of France. Now it is distributed through out the world including India.

Nature of damage : The infestation of this pest starts when the crop is still in the field and the grain is in the milk stage. This infestation continues to increase unchecked right upto the time when the harvest is waiting for threshing. The damage to the grain is always caused by the larvae which bore into grain. Once it has entered the grain, the larva eats out the kernel unseen. The infested grains are hollowed out by the larvae and filled with their excreta and webbing. Its infestation generally does not extend beyond a few inches below the surface of stored grain. Where the moisture content of the grain is high during storage, this pest is considered to be extremely serious. The maximum damage is caused during months of monsoon, *viz.*, July, August and September. According to *Pruthi* and *Singh* (1950) this pest causes more than 10% loss to various commodities.

Life history : The four developmental stages namely egg, larva, pupa and adult are found in its life cycle which have been described as below :

Egg : Female lays eggs in depressions, cracks and crevices in floors or holes in the grains, The eggs are oval in shape, about 0.5 mm in length, with both ends rounded and the surface finely scluptured. They are initially white in colour but soon change into bright red. A single female can lay upto 400 eggs which hatch in 4 days during summer and 7 days during winter.

Larva : The tiny larva crawls about a little and soon finds out a soft spot through which it enters the grain. When it is inside, closes the entry hole by a silken web and develops therein. The full grown larva is white with yellowish brown head and measures about 6 mm in length. The caterpillar moults 5 times and becomes full grown in 14-20 days which extends in winter. At this stage, the larva cuts out a circular exit hole, leaving over it just a sort of cap which can be pushed off easily by the future moth.

Pupa : Larva spins a silken cocoon inside the grain the transforms itself into a brown pupa. It measures 5 mm in length. The pupal period lasts for 7-10 days under favourable conditions.

Adult : The moth is small 6 mm in length, buff or yellowish-brown in colour with wing span of 12 mm. It has two pairs narrow pointed wings having long fringes. The head is small with black and possesses filiform antennae.

The female moth starts egg laying after 2 to 3 days of copulation. The life cycle is completed in 27-40 days depending on the environmental temperature and humidity. There are 4-5 generations in a year.

Control : Like other storage pest.

Natural enemies : *Laemophloeus minutum* and *Psocids* are the predators on the larvae and pupae of the pest.

Prevention and Control of Stored Grain Pest

There is a problem before the Indian farmers to protect their grains during storage. It has been estimated by Bindra 1975 that about 25% damage is caused from different sources to the stored grain which comes to thousand million tons of food grains. Thus we see that to become self-sufficient in food and to improve the economic condition of the country it is utmost important to check the losses caused by various agencies during storage. Grain under storage conditions suffers damage mainly from rats, insects, mites and micro-organisms. The problem of rats basically different from others which needs special attention while rest three can be tackled together.

It has been proved by experiments that damage caused by insect, mites and micro-organism depends upon the following three factors :

(i) The moisture content of the grain.

(ii) Sources, availability and amount of oxygen in the storage.

(iii) The development of temperature gradient within the stored grain.

Moisture content : For the proper development each insect and mite species requires particular humidity which in stored grains depends on the moisture content of the grain. The minimum moisture content required by some storage pests is given below :

1.	*Trogodeima granarium*	0 — 1.9%
2.	*Rhizopertha dominica*	9.0 —10%
3.	*Sitophilus oryzae*	9.5 — 11%
4.	*Tribolium castaneum*	10%
5.	*Corcyra cephalonica*	9%
6.	*Cadra cautella*	10%

It is clear from the above that the moisture content of grains is mainly below 9 to 10 per cent, then infestation of most of the pests can be avoided except that of khapra beetle. The initial moisture content before storage may be brought down by sun drying or by grain driers. The grains will be fit for storage if they are cut into pieces by teeth. Care should be taken that dried grain should always be stored in moisture proof godown otherwise it will soon adjust its moisture in accordance with atmospheric humidity which will invite the infestation of insect pests.

Availability of oxygen : Like humidity, every insect species requires a certain percentage of oxygen for its development and survival when the grains are stored in an air tight storage, the oxygen-content get reduced as it is consumed in respiration both by grain and by the insects, if present in the grain. Therefore, in the absence of minimum oxygen percentage, the development of an insect ceases and it dies. For example, khapra beetle required 16.8% oxygen content for the survival of eggs, therefore, its multiplication may be checked by lowering the oxygen content to this level.

Temperature : Temperature is the most important factor controlling the development and multiplication of insects. The insects survive and thrive within a certain limit of temperature beyond which their activity ceases. Certain insects form heat spots for their survival. The development of heat spot is due to the excessive heat produced either by insect multiplication or by microbial infection. Insect respiration produces both temperature and moisture increases, leading to favourable conditions for their multiplication. Therefore, efforts should be made to prevent the heat spots by way of cooling which may be done by forced aeration, turning of the grain and control of infestation by fumigation etc.

If these principles are kept in mind during storage the damage can be minimised upto a greater extent. Before discussing the preventive and control measures it would be worthy to know the sources of infestation during storage which are given below :

1. Some insect pests like rice weevil, pulse beetle and grain moth, come with the grain from the field it self.
2. During threshing they may infest the grain and reach un-noticed to the godown.
3. If they are present in bullock-carts, trucks, rails and other vehicles may infest the grain during transportation and reach to the godown.
4. They may be present in the old bags and infest the healthy grain when stored on them.
5. They are generally hidden in the cracks and crevices of the walls in the godown and become active when grains are stored in such godowns.
6. The adults may reach to godown by flying and larvae by moving from the neighbouring places.

Preventive Measures

1. Godowns should be cleaned well and white washed, if possible, moisture proof godown should always be preferred.
2. All the cracks, crevices and holes present in the floor, walls and ceiling of the store should be filled up with cement and levelled.
3. The walls of the store may be painted with coal-tar from the ground up to the height of 1.5 meter.
4. The dirt, broken infested grains and sweeping of the stores should be removed and burnt before new grain is stored.
5. If the cracks, crevices and holes of the godown are

suspected to harbour insects, the godown should be treated with any one of the following insecticides :

(a) Fumigation with ED/CT mixture (Killoptera) for 24 hours @ 10 lit/40 cubic meter of space.

(b) The ceiling and walls of godown may be sprayed with 0.5% malathion @ 3 lit/sq. meter.

Cleaning of Bags

1. Use of new bags as for as possible.
2. If the old bags are to be used, these should be disinfested by the following ways :

   (a) By dipping them in boiling water for about 15 minutes

   (b) By drying them in hot sun heat for about 6 hours.

   (c) By fumigating them with ED/CT mixture at the rate of 1 litre per 10 bags.

   (d) By dipping them in 1% malathion solution for 10 minutes.
3. About 20% of the room should be left free between the top layer of the bags and ceiling.

Cleaning of Grains and Precautions

1. Bullock carts, trucks, or other vehicles used for the transports of grains should be cleaned and washed preferably with phenyl water.
2. The grain should be thoroughly dried before storage, so that it does not have more than 9-10 per cent moisture.
3. The clean grain may be brought to the store direct from the threshing yard.
4. Only one kind of grain should be stored in a store as for possible.

5. If the grain is stored in bags, a layer of bhusa should be sprayed at the floor and bags should be kept 50 cm away from the walls.
6. Dunnage material (bhusa etc.) must be free from pests.
7. If the grain is already infested, it should be treated with ED/CT mixture @ 15 lit/40 q of grain before storing.
8. The grain may be stored with neem-seed kernel powder in the ratio of 100 : 1 to prevent the infestation of storage pests.

Remedial Measures : When the attack of stored grain insects is in progress in the godown, any one of the following fumigants may be used to check them :

1. ED/CT mixture @ 0.5 lit/metric tons of grain.
2. EDB ampule (3 ml) @ 1/q of grain.
3. Aluminium phosphide tablets @ I/metric ton of grain or 7 tablets per 28 cu. m. of space.

Storage Structures : Various storage structures of different capacity have been designed and recommended for storing the grain which prevent the infestation of rats, insects and micro-organisms. The most common are given below :

1. Pusa bin
2. Pant Nagar kuthla
3. Hapur bin

Pusa bin : It is a rectangular structure constructed over a pucca brick floor and can be made as per requirement. A mud platform of roughly 1.7 m × 1.2 m × 7 cm (thick) size is made and a polythene film of 1.8 m × 1.4 m size is placed over it. A 7 cm layer of mud is then applied over the film. The inner wall (11cm thick) of structure covering the four sides is then constructed. The inner layer of the mud roof of the structure is then made by using 5 cm thick mud slabs prepared earlier and placing them over a wooden frame. An area of 5 cm × 50 cm size is however, kept open in one corner to serve as a main hole. The entire structure is then covered with the polythene film. Finally, the outer layers of the walls of 11 cm thickness are erected all round the structure covering the polythene film. A 5 cm thick mud plaster is also put over the polythene film at the top. For making the structure rat proof, the outer wall of the structure may be constructed using pucca bricks. When the structure is filled with grain the main hole is finally sealed with a square piece of polythene film.

Pant nagar kuthla : Pant nagar kuthla is simple modification of regular kuthla prepared by farmers in their house. To prepare Pant Nagar kuthla a plateform of 30 inches diameter and 2 inches thickness was made of the thick dough prepared from bhusa and rawtank mud. A layer of bitumin impregnated hessian cloth (Tar felt) was pasted on it. A 2 inches thick layer of mud wall of three feet height in continuation was constructed and left for drying. A tin plate was fitted the base giving 9 "projection around to avoid rat entry. A outlet hole of 6 inches diameter was left at one side. As soon as it was dried 'Tar felt' with outlet hole of 6 inches was paisted over this wall. The tarfelt should always be in continuation of the previous layer and no gap should be left at any point. Over the Tarfelt layer, a layer of mud was plastered. Thus Pant Nagar kuthla of 60 inches height was completed till it reached to inlet hole of 12 inches diameter was also left. The lid for out-let and inlet were also constructed in the same way by sand witching 'Tar felt'. This kuthla was then dried till the relative humidity inside was 40-50 per cent.

Hapur-bins : There are various types of bins available for farmers use which may broadly categorised into two groups:

1. Indoor bins
2. Out-door bins.

Indoor bins : This group includes domestic designs of metal bins. Gharelu Thekka, Pucca kothi, welded wire mesh bin and RCC ring bin. Among these domestic designs are most popular.

Domestic designs : In this group seven different types of domestic designs of metal bins have been developed, the capacities or which are ranging from 3 to 27.5 qtls. These are indoor bins and may be kept in a room or varandah under a roof. All these bins are fabricated using either 24 gauge of 22 gauge G.P. sheets of different standard sizes. The locking facility has been provided in all types of domestic bins. The inlets are provided at the top and different types of outlets are provided at the bottom of the bins to facilitate unloading the different commodities of food grains. These bins are found to be suitable for storage of wheat, pulses and seed grains.

Out-door bins : It includes flat bottom metal bins, hopper bottom metal bins, composite bins and R.B. bins. Under this category flat bottom metal bins are more popular.

Flat bottom metal bins : These bins are for out-door use and five different capacities ranging from 20 to 50 quintels are available in this design. The model type-2 bins are fabricated using 20-G. P. sheets and model type-3 bins fabricated using 18 gauge aluminium sheets. The bins are supplied to the users either in the form of loose components, semi -assembled or fully assembled depending upon the distance of the transportation. In one design the bin can be erected on brick masonry base while in the other it can be erected on pre-fabricated elevated steel base. Since its height is not much, loading of the bin can be done manually through a simple lifting device provided. The bins are suitable for storage of wheat, paddy and maize. As aluminium is rust proof, periodical maintenance is not necessary. Its reflecting surface has an addition advantage in keeping the grain cool by radiating the heat.

6

Pest Management

The unceasing struggle between man and his insect enemies started even before the dawn of civilization. In spite of the numerous advances made by man in evolving newer and deadlier weapons to fight the war against insects, he has not succeeded in eradicating even one of the thousands of serious pests which damage his food and other agricultural products, destroy his possessions and even attack himself and injure his domestic animals.

There generally exists an uneasy truce between the insect pests and man, and this is termed as 'balance in nature. This balance is the result of two opposing phenomena, the 'biotic potential, i.e. the tremendous capacity of insects to reproduce and multiply and the 'environmental resistance' which keeps their numbers under check. The environmental resistance results in the death of adults before oviposition, in the mortality of the eggs, larvae or pupae of the insects because of desiccation, starvation, parasites, predators, diseases and other adverse environmental conditions. Even a slight slackening of any of the processes of 'environmental resistance' results in a population explosion of an insect species and the consequent epidemic.

The change in 'environmental resistance' may take place owing to a number of causes, either natural or operated by different agencies. Man is, perhaps, the single most important agent, who has, from time to time, disturbed the 'balance in nature', and this has caused numerous pest problems and pest epidemics. The nature, too, plays an important role in causing pest epidemics. Favourable conditions which reduce the natural mortality and bring about the rapid development of the insect coupled with the conditions unfavourable to the natural enemies of that insect, often result in a rapid increase of its population leading to a sudden pest out-break.

Various Theories

Once it is established that an insect is causing *economic losses,* it becomes necessary to control it.

The first principle underlying the control of an insect is its correct identification. When it is correctly identified, we can refer to the available information on the biology and habits of that insect and determine its most vulnerable stage, the appropriate time and the most suitable method or methods to control it. The knowledge and understanding of the ecological factors, both biotic and abiotic, affecting the population of an insect pest is necessary for planning the proper strategy for controlling it. A lot of attention is being paid to this aspect now and elaborate procedures and 'models' have been evolved to pinpoint the 'key factors', dominantly affecting the development and multiplication of a particular pest.

The choice of the proper method or methods of control becomes easier, when the above-mentioned information becomes available in respect of a particular pest. Some of the important methods for the control of insect pests are discussed below.

Physical and mechanical control method : This is one of the oldest methods, and has been in use since time immemorial. It includes measures, such as the collection of egg masses and other inactive stages, the removal and destruction of infested parts or whole plants, the beating of drums or tins and trenching. This method can prove to be effective during the initial stage of the pest incidence and when practised as a concerted effort by a large number of farmers in a particular area.

Cultural control method : This is a preventive method which is inexpensive and may prove effective and efficient, if employed after acquiring a thorough knowledge of the life-history and habits of a pest. Deep ploughing after harvesting a crop (to expose the hiding or resting insects), the removing and destroying of the stubble and other trash, adjusting the time of sowing (to avoid the peak-incidence period), clean cultivation, the removal of alternative wild hosts, catch crops and suitable rotations are some of the important measures included under this method of control.

Resistance of host plant : Host-plant resistance has generally been considered one of the components of the cultural method, but because of its importance, it deserves an independent status as a major method of protecting crops against insect damage. The method involves the utilization of the inherent property of certain strains or varieties of crops, of being less infested or less damaged than other strains or varieties of the same crop. Enormous economic benefits have been achieved by using this method in the case of crops, e.g. wheat, maize, cotton and alfalfa. Very little attention has so far been paid in India to the exploitation of this method. However, under the various coordinated projects on the improvement of any crop, special attention is being given to the evolvement of varieties resistant to major insect pests. It is realized that the success of this method requires the close cooperation of entomologists, geneticists and plant-breeders.

Biological control method : Practically every crop pest has its natural enemies in the form of parasites, predators and disease-causing organisms. The biological control involves a large-scale multiplication and liberation of such agents, or creating conditions under which the naturally occurring agents can act effectively. This type of control cannot be undertaken by individual farmers and has necessarily to be carried out by specialized agencies.

Some very outstanding successes have been achieved by using this method, but the method suffered a set-back owing to the large-scale and indiscriminate use of insecticides. The approach at present is to evolve methods by which, the biological and chemical methods can be integrated, so that the harmful effects of insecticides do not interfere with the activities of the natural enemies.

Legislative or regulatory control method : Legislative or regulatory method is a specific method mainly employed to prevent the introduction of pests from other countries or to prevent the spread of a pest from one area to another. The method is operated through specific regulations known as plant-quarantine laws. In India, some states have special pest acts, by which it becomes obligatory on the part of cultivators and governmental authorities to take appropriate steps to control a particular pest when it appears in an epidemic form.

Chemical control method : After the discovery of the insecticidal properties of DDT in 1939 by Paul Muller, the chemical control of insects has become most popular. As a matter of fact, the method is so popular that most of the cultivators and extension workers speak only about insecticides, whenever the question of insect control is raised. The main reason for its popularity is the spectacular and immediate results obtained by the use of such chemicals.

Hundreds of insecticides are now available to control different insect pests. These are used mainly as dusts, sprays or granules on the crops; dusts and granules can also be incorporated into the soil for the control of soil-inhabiting insects. In recent years, the soil application of some of the systemic insecticides has proved to be effective in controlling insects attacking crops in the early stages of their growth.

Some of the serious limitations of the insecticides have been highlighted in recent years. Out of these, the problems of insecticide residues on crops and other products used as food and fodder and on the pollution of environment have been agitating the minds of people in the developed and developing countries of the world. This realization has resulted in certain extremist quarters demanding the banning of the use of most of the insecticides. However, the sensible approach suggested and generally accepted is for a judicious and restricted use of insecticides. It is strongly felt that in the near future, there is no possibility of replacing the chemical method of controlling insects entirely by any other method.

Different Methods

In addition to the major methods of controlling insects described above, there are a number of other methods, which have either been employed successfully for controlling certain specific pests, or are still in the experimental and developmental stages. Increased attention is being paid now to the incorporation of such methods in the programmes of insect control. Some of these methods are : the use of attractants, repellents and anti-feeding compounds, radiations for sterilization (the well-known example of eradicating the screw-worm by using the sterile-male technique), chemosterilants, juvenile and sex hormones. These methods can in future replace or supplement the use of insecticides.

Pest Control

The guiding principles for selecting an appropriate method or methods of control should be that the method chosen must be economical, free from creating any other problem, immediately or in future, it should not harm the natural enemies of the pest and should be easy to operate and be readily available to an ordinary cultivator.

The choice of the proper method (or methods) of control becomes easier when the biology and habits of the pest are known and its most vulnerable stage has been determined. However so far the general tendency among those who are responsible for carrying out the control operations is to employ an easy method, giving quick results and disregarding other unfavourable consequences which may follow the completion of the operations.

Keeping in view the guiding principles in the selection of pest-control methods, it is now generally realized that any single approach to the problem of insect control is not feasible. Each method has its advantages as well as its disadvantages and also has limitations under a particular set of conditions. The current thinking all over the world is to suitably synthesize the use of as many methods as possible to control the insect pests effectively, economically and without any adverse after-effects. This approach is now popularly known as 'Integrated Pest Control'.

Despite a number of serious limitations from which insecticides suffer, it is felt that the use of these chemicals in the control of insect pests will continue for a long time to come and in the strategy for 'Integrated Pest Control', this method will occupy an important place as a curative method for reducing losses because of pests.

However, the emphasis is to be laid on the judicious and proper use of insecticides, so that this method may not impair the efficacy of other methods and create other problems. Keeping this point in view, some more information on insecticides and their proper use is given in the following pages.

Various Insecticides

The concept of using chemicals for controlling insect pests is not new; however, its importance and popularity increased only after the discovery of the insecticidal properties of DDT. Originally, these chemicals were classified on the basis of their mode of entry in the bodies of insects, viz. stomach poisons, contact poisons and fumigants. However, this classification has become outdated, because most of the synthetic organic insecticides act both as stomach and contact poisons and some have also the added fumigant action. At present, the generally accepted classification is based on the chemical nature of the insecticides.

Division of insecticides (based on their chemical nature)

Inorganic	*Organic*
Arsenicals Fluosilicates	

Plant origin	*Synthetic*
Nicotine sulphate Pyrethrins Rotenone Ryania	

Chlorinated hydrocarbons	*Organo-phosphates*	*Carbamates*	*Other groups*
Aldrin, BHC	Parathion	Carbaryl	DNOC
DDT, Endrin	Malathion	Isolan	Lethane
Dieldrin	Phosphamidon	Dimetilan	Thanite
Chlordane	Dimethoate	Carbofuran	
Methoxychlor	Monocrotophos		
Endosulfan	Disulfoton Phorate		
	Dichlorvos		

The inorganic insecticides were widely used before the synthetic organic insecticides were developed and marketed. At present, hardly any of these insecticides are being used in India. There is a feeling in some quarters that we can perhaps advantageously use Arsenicals against the leaf-feeding insects, especially when some parasites or predators are known to be active against them.

Insecticides of plant origin have also been relegated to the background by the synthetics and at present relatively very little quantities of insecticides, like Pyrethrins and Nicotine sulphate, are used to control insect pests of agricultural importance. The problem of residues of insecticides on food and fodder crops has emphasized the necessity of using insecticides which are less toxic to mammals or breakdown into non-toxic components in a reasonably short time. Thus the insecticides of plant origin can be usefully employed to control a number of insect pests, especially where the products are to be consumed within a short time after treatment (vegetables and fruits).

The use of synthetic organic insecticides has increased with leaps and bounds throughout the world. In India, only about 10,120 hectares of crops ware treated with pesticides in 1946-47; this area increased to about 17.4 million hectares by 1965-66 and the target for 1973-74 was 80 million hectares. The various insecticides required for controlling insects during this period is estimated to be 40,850 tonnes.

Starting with the manufacturers of BHC in 1958, India today produces technical grades of over 39 pesticides (insecticides, fungicides, rodenticides, molluscicides, nematocides and herbicides), including the following insecticides used for controlling crop pests: BHC, DDT, Malathion, Parathion (Methyl), Toxaphene, Pyrethrum extract, Nicotine sulphate, Fenitrothion, Methyl-deneton, Phosphamidon and Dimethoate.

Insecticides toxicity : Practically all insecticides are toxic to man and other animals; however, they differ in their degree of toxicity. Thus the hazards to man and his domestic animals depend on the values and nature of toxicity (oral or dermal).

The United States Environmental Protection Agency, Washington, has recognized three categories, viz. danger-poison, warning and caution on the basis of these values. Available data on some of the common insecticides are presented in Table.

Table. Acute Oral LD50 Values of Some Common Insecticides for White Rats

Insecticides	*Oral LD_{50} (mg/kg)*		*Remarks**
	Males	*Females*	
(A) Chlorinated hydrocarbons			
Aldrin	39	60	Toxic
Benzene hexach-loride (BHC)	1,250	—	Less toxic, sax not mentioned
Chlordane	335	430	Toxic
DDT	113	118	Less toxics
Dieldrin	46	46	Toxic
Endosulfan	43	18	Toxic
Endrin	17.8	7.5	Highly toxic
Heptachlor	100	162	Toxic
Lindane (gamma BHC)	88	91	Toxic
Toxaphene	90	80	Toxic

Contd...

Insecticides	*Oral LD_{50} (mg/kg)* Males	Females	*Remarks**
(B) Organo phosphates			
Aldicarb	0.93		Highly toxic, sex not mentioned
Diaxinon	108	76	Toxic
Dichlorvos	80	56	Toxic
Dimethoate	215	—	Toxic
Disulfoton	6.8	2.3	Highly toxic
Fenitrothion	740	—	Less toxic, sex not mentioned
Malathion	1,375	1,000	Less toxic
Methyl parathion	14	24	Highly toxic
Mevinphos	6.1	3.7	Highly toxic
Monocrotophos	17.5	20	Highly toxic
Parathion	13	3.6	Highly toxic
Phorate	2.3	1.1	Highly toxic
Phosphamidon	23.5	23.5	Highly toxic
(C) Carbamates			
Carbofuran	11	—	Highly toxic, sex not mentioned
Carbaryl	850	500	Less toxic
Zectran	37	25	Toxic
(D) Plant origin			
Nicotine sulphate	—	83	Highly toxic

Contd...

Insecticides	*Oral LD_{50} (mg/kg)* *Males*	*Females*	*Remarks**
Pyrethrins	1,500	—	Less toxic, sex not mentioned
Pyrethrum	1,870	820	Less toxic
Rotenone	50-75		Less toxic, sex not mentioned
Ryania	1,200	—	Less toxic, sex not mentioned
(E) ***Inorganic***			
Calcium arsenate	—	298	Less toxic
Lead arsenate	—	1,050	Less toxic
Paris green	—	100	Less toxic

Use of insecticides and precautionary measures : The improper, excessive and careless use of insecticides may prove to be injurious to man and his domesticated animals. Proper safeguards are, therefore, necessary to protect the persons handling the insecticides, the crops to which they are applied, the consumers who have to use the produce of the treated crops, and the domesticated animals which feed on such produce, the pollinating insects, the parasites and predators of insects, the fish, birds and other wildlife preying upon insects, likely to be affected by coming into direct or indirect contact with the insecticides applied. Also, the pollution of air, soil and water is to be avoided as far as possible.

Some of the basic precautions to be taken while handling, using and storing insecticides are as follows:

1. Keep the insecticides in closed, properly labelled containers in a dry and cool place, away from food and fodder, and in places where children and animals cannot reach them.
2. Use the insecticides according to the instructions given on the containers, and adhere to the dosages recommended.
3. Persons handling insecticides should avoid the contact of the insecticides with their skin (specially that of the insecticides listed as toxic and highly toxic) and the inhalation of dusts, vapours or mists. The minimum precautions of wearing rubber gloves and covering the eyes and nose should be taken. The operators must not smoke, eat or drink anything while applying insecticides. After finishing their work, they should take bath, wash their hands and face with soap thoroughly and change their clothes. The clothes worn by them during the operations should be washed properly.

Application equipment for insecticides : The proper choice of the equipment to apply insecticides depends on a number of factors, the first of which is the formulation to be used. Insecticides can be applied as dusts, sprays (as high-volume and low-volume sprays, aerosols, mists or fog), or as granules, and the application machinery has to be chosen accordingly. The next important factor to be considered is the size of the area to be treated. For small holdings, hand-operated applicators may be suitable, whereas for large holdings power-operated machines will be needed. Similarly, due consideration is to be given to the crop to be treated and the pest to be controlled.

Different kinds of dusters and sprayers available in the market are listed below.

Dust particles : Manually-operated dusters for small

holdings. These are plunger-type bellows duster, knapsack duster, and crank or rotary duster (very commonly used in India).

Dust power : These dusters are used to cover large areas and for quick coverage. These are engine and traction-operated.

To obtain good results and for proper coverage, a duster should produce a continuous and uniform cloud of the insecticide. The dust should be dry and fine enough not to clog the hopper or lump inside it. The duster should be light and portable and should be easily repaired even by the village blacksmith.

Different Sprayers

Sprayers used by man : These are hand-sprayers or atomiser (for treating a few plants only), stirrup pump or sprayers, slide-pump sprayer, knapsack sprayer (pneumatic or battery-operated), compressed-air sprayer, rocker sprayer, foot or pedal sprayer and wheelbarrow sprayer. Pneumatic knapsack and foot or pedal-pump sprayers are commonly used by the Indian farmers with small holdings.

The power-operated sprayers include the equipment operated with gasoline engines, such as stretcher sprayers, wheelbarrow sprayers, boom sprayers and self-propelled high-clearance sprayers. Power-sprayers are generally available with village panchayats or zila parishads, from which the farmers can obtain them on hire, whenever necessary.

In addition, there are special types of machines for producing mists, fogs and aerosols which have limited use. Recently, granule applicators have been marketed for applying granular insecticides.

The proper care of application equipment is of utmost importance. The machines should be kept in good working order even when not required immediately. They should be cleaned thoroughly, oiled and greased before storing after use. In the case of sprayers, insecticides should be completely removed from the tank and the delivery tube and properly dried; their nozzles should be cleaned and any leaks should be immediately repaired.

Use, how much insecticides: For controlling the insect pests of crops, the usual practice is to indicate the dosage, i.e. the quantity of the active ingredient to be applied per unit area (acre or hectare). However, the dosage may have to be varied, depending on the type and the stage of the crop. Another method is to indicate the concentration and the quantity of an insecticide to be dustered or sprayed. Here also, the quantity of the dust or of the spray liquid has to be adjusted according to the type and the growth stage of the crop and according to the application equipment available. In the following pages, concentrations of insecticides have been recommended, keeping in view the general use of high-volumes prayers, by which the plants may be thoroughly covered up to the run-off stage. The quantity of the spray may, therefore, have to be adjusted accordingly.

Effect on Crops

The crops grown in India are attacked by a large number of different pests, some of which are specific to a particular crop or feed on related species of a particular family, or to crops grown in a particular season. However, there are other insects, such as termites, army worms, cutworms, grasshoppers, and locusts which attack a large variety of crops in different seasons. Out of them, locusts occupy a unique position because of their swarming habit and migrations to distant areas irrespective of international boundaries. An account of this group of insects, specially the desert locust, which is regarded as 'Farmers' Enemy No. 1', is given separately.

Important locusts : India has been subjected, from time immemorial, to locust invasions which cause immense damage to crops and trees, sometimes resulting in severe fodder famines and considerable loss of livestock.

General features : Locusts are closely allied to short-horned grasshoppers from which they differ markedly in certain respects. Both are anatomically very similar, feed on vegetations, lay eggs in the soil and pass through several instars before they become adults. Whereas grasshoppers feed on specific crops only, locusts devour almost all plants and vegetation, and even feed upon dead and dying locusts, i.e. they are omnivorous. Unlike grasshoppers, locusts exist in two phases: (*i*) the solitary phase, and *(ii)* the gregarious phase. In the solitary phase, they are inactive, live and breed as scattered individuals, and are relatively harmless. In the gregarious phase, they are very active, acquire the habit of congregating and forming bands, breed rapidly, and in the adult stage fly long distance during migration even overseas. Climate, weather, soil, vegetation and their mass destruction organized by man cause the transformation of the gregarious phase into the solitary phase.

Types of locust : There are eleven main species of locust, of which three, viz. the Bombay locust, the migratory locust and the desert locust, are found in India. The first two species live and breed in the country and are found scattered all over western India from Rajasthan in the north to Karnataka and Tamil Nadu in the south. They rarely form swarms. They mainly feed on plants of the grass family and occasionally when they become gregarious, do considerable damage to *jowar, bajra* and other millets.

Desert locust distribution : The desert locust is the most destructive international pest, its distribution extending from India in the east to Africa in the west, passing through Pakistan, Afghanistan, Iran, Iraq, southern USSR, Arabia and in the north, east and west of Africa. In this vast region, the locust lives and breeds endemically, forming swarms and undertaking long flights to distant places. The locust invasions occur in cycles, the duration of a cycle as well as of the period between two cycles being variable.

Different phases : In the solitary phase, the locust is more or less harmless, whereas in the gregarious phase, when its hoppers congregate and march in bands, destroying all vegetation in their path and give rise to swarms, it is very destructive. The hoppers of the solitary phase are green throughout their life, but the hoppers of the gregaric is phase are black in the first two stages of development and their dark colour turns greenish yellow and orange in the later stages. The adult locust of the solitary phase remains grey, but the adult is pink at first, but turns grey and yellow as it matures.

Breeding time : There are generally two breeding seasons in a year: (i) winter-spring, and (ii) summer-monsoon. In regions where the rainfall is mainly received in winter and early spring, e.g. in south-east Arabia, southern Iran and Baluchistan, the locust breeds in spring. In India and other countries where rainfall is received during the summer months, breeding takes place at that time of the year. The swarms produced during winter and spring in the Middle East and Baluchistan usually fly eastwards into Pakistan and India, in summer. With the advent of the monsoon in this area, they lay eggs from June to September. The swarms resulting from this breeding return during autumn to the areas of winter rainfall when conditions in the summer-rainfall area become dry and unsuitable for breeding. A number of swarms developed in the country fly also to the north, south and east, and thus invade all parts of India and damage the *kharif* crops.

Sometimes, breeding takes place in winter, specially in Punjab and Haryana when they damage the *rabi* crops as well.

History of growth : The female of the gregarious phase lays eggs in clusters of 40-120 in a hole in moist soil to a depth of 75 to 150 mm and covers them with a liquid material that soon hardens to form a protective covering. Young hoppers hatch from the eggs in two to three weeks in summer, and in three to four weeks in autumn and spring. There are five stages in the development of hoppers. At each stage they cast off the skin every sixth or seventh day as they progressively become bigger. As long as they are hoppers, they have only rudimentary wings. After the fifth moulting, they develop wings from the wing-pads present in the late hopper stages. During summer and monsoon, the hoppers complete their development in four to five weeks; and during autumn and spring in six to eight weeks. As the hoppers moult, they undergo a change in colour and size. In early stages, they are almost wholly black, and in the later stages, they are black, with orange and yellow patterns. When they acquire wings, they are pink and ultimately turn yellow, when fully mature.

Both hoppers and adults feed on all types of vegetation, though there are some plants which they would not touch, or eat only with reluctance.

Locust cycles and biotic theory of periodicity : The theory has been recently propounded to explain the causes of periodic locust outbreaks. According to this theory, the actually adverse climatic conditions of the desert are, in fact, favourable to the locust, as they adversely affect its enemies (specially cold-blooded reptiles), which are responsible for not allowing the locust to breed outside the desert, where the conditions are otherwise very favourable. The desert environment becomes more inhospitable to the enemies of the locust during the minima of the sun-spot cycle (periods of drier and more intense heat), leading to a reduction in the population of the locust enemies which consequently leads to an increase in locust population and swarm formation. During the maxima of the sun-spot cycle (period of less dry and less intense heat) on the other hand, the desert environment becomes favourable to the enemies of locusts which migrate to the interior of the desert from the periphery and bring down the locust population, thus breaking the locust cycle.

This biotic theory has made it possible to visualize the possibilities of checking the start of the locust cycle in the early stages by creating suitable ecological niches in the locust-breeding areas, so that their enemies can migrate and survive the adverse conditions in such areas and continue to keep down the locust population.

Measures of control : There are two main methods of locust control, viz. *(i)* the use of poison baits, and *(ii)* dusting and spraying with insecticides. Poison-baiting, though cheap, is not very practicable, dusting and spraying have been found to be very effective. In the egg stage ploughing or flooding the land has been practised in the past. This method is not very effective inasmuch as it destroys only about 30 per cent of the eggs and renders subsequent control measures with insecticides difficult. Spraying the egg-infested, uncultivated land with Aldrin is very effective, as it leaves a poisonous film on the surface with which the emerging hoppers come into contact and die. Once the hoppers have emerged, they can be effectively controlled by dusting or spraying or both, as the hopper stage is most vulnerable. For dusting, BHC is the most effective insecticide. For young hoppers, 3 to 5% dust is sufficient to kill them; for advanced stage, 5 to 7% dust is necessary; and adults need 10% BHC. Dusting or spraying may be done on the hoppers directly or it may be done on strips of ground, two to three metres wide, in the path of the advancing hoppers. Another method is to dig trenches 45 to 75 cm deep and 30 to 60 cm wide, depending on the size of the hoppers, in front of the moving hoppers to which they may be driven and buried alive. The dusting or burning of bushes in which hoppers rest can also be resorted to.

Adult locusts, flying in swarms, are most difficult to control. If they alight in any field or in a piece of ground try to kill as many of them as possible, by beating or crushing, specially in the morning when they are lazy. Burning them by setting fire to dry bushes on wastelands, or by flame throwers, if they have settled on green bushes, hedges or trees if useful. This method works well if the locusts have settled for resting, mating or egg-laying. Poison baits-spread at the rate of 20 to 50 kg per hectare are useful if locusts have settled to feed in cropped feeds. If the swarm consists of pink locusts, dust them with 10% BHC. Aerial spraying of the flying swarms and ground hoppers has to be resorted to with Aldrin solution in oil in the case of very serious infestation. However, this cannot be done when the swarms and hoppers are in the populated areas.

Despite the discovery of a large number of potent insecticides, which can be used against locust, both from ground and air, the farmer has been finding himself to be only a helpless onlooker once the locust swarm settles in his field. Though part of the swarm may be killed, he invariably loses his crop.

Situation has materially changed recently after the discovery of the anti-feeding property of the *neem* kernels. It has been found that 0.1 per cent suspension *of neem* kernels, sprayed on the crop affords protection against locust feeding for about 14 days. The cultivator can locally collect and preserve the *neem* seeds till required. By spraying his crops with a *neem*-kernel suspension, when the invasion of his fields, by hopper bands or swarms is imminent a cultivator can now ensure the safety of his crops. Simultaneously, the control operations can be carried out by adopting suitable methods.

Organizational structure : As the locust is an international pest, control measures are effective only when they are properly organized at the national and international levels. Since 1951, the FAO has taken good interest in the desert-locust control by periodically appraising the overall situation, recommending suitable control techniques and organizational patterns, and by coordinating and contributing to the carrying out of control campaigns in certain key areas. More recently, it has been organizing international campaigns annually in the Arabian Peninsula with the assistance of a number of countries.

In India, the Central Government has assumed direct responsibility for controlling locusts in the desert areas of Rajasthan, Punjab, Haryana and Gujarat. This area is divided into four circles with headquarters at Jodhpur, Barmer, Bikaner and Palampur, each of which is subdivided into zones and the zones are further divided into outposts. In each zone and at each outpost, technical staff, control equipment, vehicles, etc. are maintained. Wireless-sets are used for transmitting messages and aeroplanes are available for aerial spraying and dusting.

In the cultivated or non-desert areas, the responsibility for controlling locust rests with the state governments. The State Plant Protection Organization provides the necessary technical staff, equipment, vehicles, etc. The District Magistrate is in charge of locust-control operations, ill carrying out which, he is helped by the State Plant. Protection Officer and the district staff and extension workers of the Revenue and Agricultural Departments.

In locust control, the exact information on mating, egg-laying, hopper concentration, swarm movements, direction of their flights, the size of the locusts, their colour, etc. is of utmost importance for the rapid transport of the necessary equipment and insecticides, the posting of the technical staff, the supply of labour, etc. All people, both officials and non-officials, can therefore, render help by the prompt transmission of the information of the kind mentioned above; they may be having or may have heard to the nearest authorities concerned, e.g. any officer of the Revenue or Agricultural Department or of Community Projects or National Extension Service.

About other major pests of different crops and their control, information in a condensed form is given in the following pages.

Various Diseases

How to define : A plant disease is a physiological or structural abnormality that is deleterious to the plant or to any of its parts or products, or reduces their economic value. It expresses itself in a definite series of symptoms.

Nature of disease : Plant diseases may be temporary or permanent, depending on the speed of recovery. Further, a disease may be localized, involving only a limited part of a plant, or it may be systemic when its infection is generalized and spreads throughout the plant body.

Causes of disease : Plant diseases may be caused by micro-organisms, including fungi, bacteria, viruses, mycoplasmas, etc. or may be incited by physiological causes including high or low temperatures, lack or excess of soil moisture and aeration, deficiency or excess of plant nutrients, soil acidity or alkalinity, etc. The micro-organisms obtain their food either by breaking down dead plant and animal remains (saprophytes) or by attacking living plants and animals (parasites). In order to obtain nutrients, the parasitic organisms excrete enzymes or toxins and kill the cells of the tissues of the host plant, as a result of which either the whole plant or a part of it is damaged or killed, or considerable disturbance takes place in its normal metabolic processes.

How to reproduce : Micro-organisms are reproduced and disseminated mainly by spores. Wind, water, seed, diseased plants, cuttings and tubers, animals, men and insects, and soil help to disseminate diseases. Virus diseases are spread as the result of direct contact, through seeds, diseased cuttings and grafts, and more commonly by insect and other vectors. Most of the fungi and bacteria errupt or ooze from the surface after the host tissues have been damaged, whereas the viruses work altogether internally.

Therefore, the control of plant diseases usually lies in

preventing infection, as it is very difficult to kill the fungi or bacteria or to inactivate the viruses inside the host plants. However, as the result of recent studies made in the development of systemic fungicides, it appears that in the near future the diseased plants may even be cured. The revolvement of disease-resistant varieties is another important and cheap method of controlling diseases. However, this method of control has its limitations because of the appearance of new races of the pathogens from time to time.

Disease borne by seed itself : In the case of seed-borne diseases, the pathogens are carried either on the surface of the seed or within it. When a pathogen is external, it may be destroyed by treating the seed with a chemical, e.g. formalin, TMTD, copper carbonate, Captan. organo-mercurials, e.g. Agrosan and Ceresan. Except Formalin, all chemicals are available in the dust form. The recommended dust is thoroughly mixed with the seed, preferably in a rotating-drum, so as to uniformly cover the seed surface with a layer of it. While using formalin, dilute it with an equal quantity of water and spray it over a thinly spread layer of grains. The grains are then heaped and kept covered with a tarpaulin overnight. For internally seed-borne diseases, hot-water treatment and solar treatment are very efficacious.

For the treatment of wheat seed with hot water to control loose smut, the seed is soaked in ordinary water for four to five hours and then dipped into hot water at 49° C for one minute and then immersed again into hot water at 54° C for ten minutes. In the case of solar treatment, which is carried out on a bright sunny day during the summer months, the seed is soaked in water for 4 hours from 8 a.m. to 12 noon. It is then taken out and immediately spread in a thin layer on sheets of cloth to dry in the sun. Recently, systemic organic compounds have also been shown to be effective in eradicating both externally and internally seed-borne infections.

Disease borne by soil : In the case of soil-borne diseases which attack the seeds or seedlings, the use of soil disinfectants or sterilizers is helpful in destroying the pathogens. The soil treatment involves the use of chemicals formaldehyde, Captan, Thiram, Zineb, organo-mercurials, PCNB, ethylene dibromide, Vapam, etc. They are used to drench the soil, or are broadcast on it, or are applied to it in furrows, or are used to fumigate it.

Disease born by air : In the case of diseases caused by air-borne pathogens, the application of chemicals to the foliage is the most effective way of controlling them. Though many fungal and bacterial diseases can be controlled in this way, chemicals have not yet proved very successful in controlling virus diseases. In their case, protective measures have to be aimed at controlling the insect vectors by spraying insecticides on the host plants or by using disease-free material for propagation.

Fungicides : Copper, sulphur, organo-mercury and organo-sulphur compounds and antibiotics are some of the important fungicides. Recently, some systemic organic compounds have also been developed. The choice of fungicidal compound depends on the nature of the disease and on the susceptibility of the crop. They are available as dusts or wettable powders and sometimes in liquid form. The method of applying them depends on the availability of water, suitable machinery, the nature of the crop to be treated, the area to be covered and the cost of labour.

Fungicides with copper : Bordeaux and Burgundy mixtures, and the dispersible forms of cuprous oxide, copper oxychloride, and basic carbonate are among the important copper fungicides. Bordeaux mixture is prepared by mixing solutions of copper sulphate and quicklime in certain proportions. For preparing 5:5:50 Bordeaux mixture, 2 kg of copper sulphate and 2 kg of quicklime are dissolved in 225 litres of water. At first, the stock solutions of the two constituents (1/2 kg in 4 litres of water) are prepared. Copper sulphate is dissolved by placing it in a piece of gunny sacking and then suspending it in water in an earthen or wooden vessel.

Lime is first slaked with a small quantity of water and is made into a thin paste before the remaining water is added to it. The two stock solutions are mixed together just before use by pouring the copper sulphate solution into the lime solution. The final reaction of the mixture should be either neutral or slightly alkaline. The reaction can be tested with litmus paper, or by dipping a bright iron nail into the mixture. The litmus paper must not turn red and the nail must remain bright, i.e. without any copper deposit on it. Bordeaux mixture deteriorates on standing; hence only fresh preparations should be used. However, this deterioration can be prevented appreciably by adding cane-sugar to the mixture at the rate of 60 grammes per 225 litres.

Burgundy mixture is a variation of Bordeaux mixture, in which lime is replaced by sodium carbonate (washing-soda). The method of preparation is just the same as for Bordeaux mixture. The usual strength of the solution is 4—6¼—50. There are a number of proprietary copper fungicides available in the market, e.g. Perenox, Persian, Blitox, Cuprokyt, Cuprosan, Fytolan, which are as efficacious as, and in some cases better than, Bordeaux mixture. The proprietary products also contain dispersing and wetting agents which make them more effective.

Fungicides with sulphur : The important sulphur fungicides in use are sulphur powder, wettable sulphur, lime-sulphur and organo-sulphur. Sulphur-dust is a very cheap fungicide, but its efficacy depends upon the size of the particles which should pass through a sieve of 250 to 300 meshes per inch. Wettable sulphur is used on crops which are liable to suffer from lime-sulphur scorching. The lime-sulphur solution is prepared as under :

Water (225 litres) is first boiled in a vessel of iron, tin or zinc (but not of copper) and then quicklime (20 kg) and sulphur (55 kg) are added to the boiling water. The mixture is stirred and boiled for one hour. The resultant liquid is strained and its density is determined with a hydrometer ; enough water is added to bring the specific gravity to 1.005 or 1.010 (see table, p. 300) before use. Proprietary products, e.g. Sulfinette and Limsul, are available in powder form. On dilution, they are ready for use as the lime-sulphur spray. The lime-sulphur spray should not be used on tender foliage, as it is likely to scorch it. A number of organo-sulphur compounds, e.g. thiram and dithiocarbamates, are extremely safe to use on most of the plants with tender and sensitive foliage. Tulisan, Zineb and Ziram are some of the proprietary products, based on organo-sulphur compounds.

Dilution Table for Lime-sulphur

Density (Degree Baume)	*Gallons of water to be added to one gallon of concentrated solution to give :*	
	Sp. gr. 1.005	*Sp. gr. 1.010*
28	22	46
29	23	48
30	24	50
31	26	52
32	27	54
33	28	57
34	29	59
35	30	61
36	31	63

(a) *The Organo-mercurials* : These have proved to be effective in controlling seed-borne and some soil-borne diseases. They are applied as dusts in extremely small doses to the seed before sowing. They have no harmful effect on germination. Rather, they prevent emergence losses caused by soil-borne pathogens and result in a better stand of the crop. There are a number of organo-mercury preparations, e.g. Agrosan GN, Ceresan, Semesan and Parrugen. They are, however, poisonous to human beings and animals. Hence great care is necessary in handling these fungicides and the seeds treated with them. Some vegetable seeds and other seeds, e.g. those of linseed, are non-tolerant of mercury. In the recent years certain chemicals, such as Captan, Thiram, Busan-72 and Dithane M-45, and some systemic fungicides have been developed. They may be used for seed treatment instead of the mercurials.

(b) *The systemic Fungicides* : Systemic fungicides can move mainly upwards only within the plant and kill the pathogen away from the point of application. They can be applied as seed or soil treatment, and as sprays. Some of the important systemic fungicides are : oxathin derivatives (Plantvax and Vitavax), Benlate, Demosan and Bavistin.

(c) *The antibiotics* : Antibiotics are mainly used against bacterial plant pathogens. Some of the promising antibiotics are : streptomycin sulphate, streptocycline, terramycin, agrimycin-100 (a mixture of streptomycin and tetracycline), agrimycin-500 (containing copper), etc.

Cultural control : Cultural practices, such as crop rotation, mixed crop ping, adjustment of the date of sowing, depth of sowing and field sanitation through the removal of diseased debris, weeds, alternate and collateral hosts are very effective in controlling plant diseases. A strict enforcement of quarantine, especially in the case of imported plants, helps to prevent the introduction of new diseases into a country.

As virus diseases are difficult to control or cure with chemicals, the use of disease-free propagating material is very important. For this purpose, it is necessary to establish disease-free nurseries at isolated places.

Control measures against some of the important diseases, together with their casual organisms and symptoms, are described in the following pages.

Plant Ailments

Though plant parasitic nematodes often cause severe damage to cultivated crops, their role as a limiting factor in agricultural production has not been generally appreciated. These microscopic organisms, sometimes also called thread worms, eelworms, round worms, etc. mostly inhabit the soil in association with the roots of plants. A few species feed 0:1 the aerial parts of plants too, such as stems, leaves, ear heads, grains, etc. Unlike insects and other disease-causing agents like bacteria, fungi and viruses, nematode populations increase rather slowly, and the diseases caused by them are, generally, not of sudden epidemic type resulting in obvious destruction of the crop over vast areas in a matter of days, but cause slow decline in yields spreading over years. Symptoms of nematode diseases may not often be clear-cut and easy to pin-point, and the disease may itself be the result of interaction between nematodes and other pathogens like bacteria, fungi and viruses.

Many estimates have been made of crop losses caused by pests and diseases. However, many of these are often questioned as being unrealistic. Assessment of crop losses due to nematodes is even more difficult as compared to insects for want of precise techniques and adequate survey data. Some workers have attempted to express these losses in terms of money. In the USA alone, the estimated annual loss in cultivated crops due to plant nematodes was considered to range between 250 and 500 m dollars till a few years ago. A Committee on Crop Losses recently appointed by the Society of Nematologists, USA, has undertaken a detailed study on the subject. On the basis of an average 6 per cent loss in field crops, 12 per cent loss in fruit and nut crops, 11 per cent in vegetable crops and 10 per cent in ornamental crops, they estimated the annual losses in farm value suffered by that country to a staggering total of 1,590.7 m dollars. The factors considered in making these estimates included reduced yield, reduced market quality owing to nematode damage, increased yield or quality or both after chemical treatment and cost of chemical control measures. Data from experiments involving extreme damage levels by high nematode populations and greatly accelerated growth rates after successful control were discounted. An assessment in England and Wales, some time ago, indicated that the annual loss of potatoes caused by the golden nematode of potatoes alone was of the order of 200,000 tonnes representing £2,000,000.

In India, some attempts have been made recently to estimate losses due to nematodes in a few crops. Annual losses as high as Rs 40,000,000 in wheat and Rs 30,000,000 in barley due to the *molya* disease caused by *Heterodera avenue* in Rajasthan alone are considered common and reasonably correct. The ear-cockle nematode of wheat has been estimated to cause an annual loss of over Rs 75,000,000 even at a low infestation level of less than 1 per cent. Another disease, a decline of coffee in South India, caused by a root lesion nematode *Pratylenchus Coffee* has been estimated to cost that industry rupees 20,000,000 per year. It is rather premature to attempt estimation of losses in other crops for want of sufficient information.

Nematode pathogens : Nematode parasites of higher plants are by no means the only nematode fauna of the earth. There are numerous free living forms in the soil, besides parasites inside the bodies of man and animals. Countless numbers inhabit the seas also. Even so, plant parasitic nematodes are so wide-spread and universal in distribution that no living plant is likely to be free from their attack. All cultivated crops are usually attacked by one or more species of these nematodes. The actual damage done to the host plant, however, would depend on the population, the age of the plant at the time of infestation and several other ecological factors.

Plant nematodes are mostly microscopic, generally ranging from 0.5 to 1.0 mm in length. Like insects, they vary in size, appearance and habits. Typically they are cylindrical, the body tapering towards the head and tail-ends. A departure from this typical slender shape is observed in mature females of many species which assume swollen pouch-like, lemon, pear, kidney or other irregular shapes while the male remains slender. Perhaps the most outstanding diagnostic feature of plant parasitic nematodes is that they all have at the mouth end a stylet or spear, which is a sharp pointed tube like a doctor's hypodermic needle. It is worked back and forth into plant tissue with the help of three sets of muscles attached to its base. The plant sap is sucked into the alimentary canal through the stylet tube by the vacuum created by the alternate contraction and relaxation of a median valve situated posterior to the stylet. At the same time, the nematode also injects a small quantity of its oesophageal gland secretion containing digestive enzymes into the plant tissue. The reaction of the plant and the expression of various disease symptoms are attributable to some toxic substances thus injected and to depletion of plant nutrients while feeding. Nematodes have six stages in the life cycle, the egg, four larval stages and the adult. One cycle may be completed in a few days or may require several years depending upon a great variety of conditions and circumstances.

The nematode parasites of plants can be classified in a number of ways on the basis of feeding habits, parts of plant attacked, the type of reproduction, etc. A simple classification based on their parasitic habit is adopted below :

I. *The ectoparasites* : These nematodes are external feeders and normally do not enter the plant tissue. They feed on the cells of small feeder roots near the root cap and may cause death of cells near the feeding site resulting in collapse of tissues. Spiral nema todes *(Helicotylenchus, Rotylenchus)*, lance nematodes *(Hoplolaimus)*, ring nematodes *(Criconemoides, Hemicriconemoides)*, stunt nematodes (*Tylenchorhynchus)* etc., are good examples. Some of these ectoparasites may cause small galls on the roots, as for example, sheath nematodes *(Hemicycliophora)* and dagger nematodes *(Xiphinema)*. Besides the root feeding forms mentioned above, certain species of foliar nematodes *(Aphelenchoides)* also feed ectoparasitically on the leaves and flower buds. Deformed leaves and fruits or a reduction of these parts may result from heavy infestation.

II. *Semi-endoparasites or exposed endoparasites* : The citrus nematode *(Tylenchulus semipenetrans)* and the reni-form nematode *(Rotylenchulus reniformis)* are two good examples of this category, both widely distributed in India. The adult females are saccate or swollen with the head and neck region alone embedded in the roots and the hind part of the body protruding outside the root tissues. Both species cause appreciable damage to their hosts and are of great economic importance to India. While the citrus nematode is mostly specific to citrus, the reniform nematode infests a variety of crops like millets, pulses, cotton, castor, vegetables, chillies, potato etc., and is one of the most harmful species in the country.

III. *The endoparasites* : These nematodes are completely internal parasites and include some of the most destructive crop pests known to man.

(i) *Nematodes of root-knot (Meloidogyne spp.)* : These are the most common plant parasitic nematodes in

India and their infestations are easily recognised by the characteristic root-galls they cause. Vegetable crops like tomato, brinjal, bhindi, cucurbits, crucifers, pulses, potato, sugarcane, banana, citrus etc., are all highly susceptible hosts. Even cereal and millet crops, which are generally not favoured, are attacked under certain conditions and suffer considerable damage. Within the root-galls are harboured the highly swollen or flask-shaped adult female nematodes which are sedentary in habit, incapable of movement within the roots. More than half a dozen species of root-knot nematodes are known to occur in India infesting various crops. *M. incognita* and *M. javanica* are the most common species.

(ii) *Nemotodes regarding cryst-forming (Heterodera spp., Globodera spp.)* : These are also sedentary endoparasites and probably the most highly specialised among the plant nematodes, because of the special adaptation for survival which they possess, namely, the cyst stage in the life-cycle. Eggs are generally not laid outside, but retained within the female body wall itself which eventually gets hardened into the brown, leathery cyst. The cysts enclosing the eggs get dislodged into the soil and are very resistant to desiccation and other adverse conditions. It is mainly this adaptation which makes the cyst-forming nematode the most difficult to control. *G. rostochiensis* and *G. pallida* on potatoes in the Nilgiris, *H. avende* on wheat and barley in north-western India, *H. cajani* on pulses and *H. zeae* on maize are good examples.

(iii) *Nemotodes regarding root-lesion (Pratylenchus, Hirschmanniella, Radopholus)* : These are internal root parasites whose adults are migratory in habit and move about through the cortical tissues of the root. Extensive injuries in the form of root lesions are generally caused, often with severe symptoms. Host plants include many important cultivated crops such as rice, vegetables, millets, pulses, cotton, sugarcane, bananas, citrus, coffee and many others. The root-lesion nematodes *Pratylenchus thornei, P. zeae, P. delattrei, P. coffeae* etc., on a variety of crops; the burrowing nematode *Radopholus similis* on banana, peppervine and coconut in South India ; and the rice-root nematode *Hirschmanniella oryzae* are more important species.

(iv) *Parasites in flower and seed :* Under this category are included the white tip nematode of rice, *Aphelenchoides besseyi,* which is a seed-transmitted nematode causing distortion and chaffiness in rice grains, and the wheat gall nematode, *Anguina tritici* which produces galls in the flowering parts of the host plant. These two species are known to cause wide-spread damage to rice and wheat respectively in India.

(v) *Nematodes regarding foliar (Aphelenchoides spp.) :* These live as internal parasites of buds and foliage of plants like chrysanthemum, strawberry, etc., and may cause distortion and crinkling of foliage, and leaf spots. At least one species *A. ritzemabosi* is known to infest ornamental crops in Simla and other hill areas.

(vi) *Nemotodes regarding stem and bulb (Ditylenchus spp.):* These occur more commonly in temperate regions of North America and Europe. *D. dipsaci* infests onion, garlic, alfalfa, clover, bulbs, ornamentals etc., in these countries. *D. destructor* causes dry rot of potatoes. There is no published record to indicate the presence of either species in India.

Plant injury : Plant parasitic nematodes are obligate parasites. During feeding in or on the roots, they penetrate cells and inject their salivary fluid into the host tissues. It is apparent that a number of digestive enzymes like carbohydrates and proteases are released during the process of feeding resulting in hydrolysis of host components and leading to altered host metabolism. Feeding on plant cells by nematodes may deprive the host of nutrients and water necessary for its growth. The conductive tissues responsible for the translocation of nutrients and water to the tops are often blocked.

Metabolites vital for the growth and maintenance of host tissues are often blocked at the site of infestation. Feeding injury will thus impair the efficient functioning of the organs concerned. The impact of mechanical injury to the host tissue brought about during the process of feeding or penetration by enormous number of nematodes has also significant effect on plant growth.

Many plant parasitic nematodes, e.g. *Meloidogyne* spp., *Rotylenchulus reniformis* etc., may be directly responsible for poor germination and thin seedling stand due to pre and post-emergence seedling damage under field conditions. Root-knot and reniform nematodes are also known to interfere with rhizobial nodulation in leguminous crops leading to reduction in the number and size of nodules and consequently reduced nitrogen fixation. Often nematode injury leaves avenues for the entry of other secondary organisms. The association of plant parasitic nematodes with other microorganisms, like fungi and bacteria in disease complexes, is now a well recognised phenomenon and very often such association increases the severity of the disease. Further, the altered metabolism induced by nematode parasitism makes the plant vulnerable to the attack of otherwise weak pathogens. The discovery of some of the plant parasitic nematodes to be the vectors of soil-borne virus diseases indicates further potentiality of these small animals to cause injury to crops.

Damage and symptoms : The symptoms of nematode injury to host plants will vary with the species of nematodes involved, initial population density, age of the plants and many other ecological factors. Loss of vigour of individual plants, patches of stunted plants, general decline followed by reduction in crop yield etc., are some of the common symptoms. Field symptoms are often of a generalised type, unspecific, giving an impression that the crop is suffering from shortage of water and fertilizers even though both these factors are available in plenty in the soil. General chlorosis, die-back, leaf drop, small-sized foliage, dead buds, leaf distortions, under-sized and premature ripening of fruits, root galls, root rot, root lesions, excessive proliferation of roots or devitalized root tips are some of the symptoms of nematode injury to the hosts. Since symptoms on host due to nematode infestation are similar to those caused by other pathogens like bacteria, fungi, etc., laboratory examination of soil and root is necessary to make sure of the role of the nematode in the manifestation of the disease.

Host and parasite : Among all the known pathogenic entities, the kind of relationships that exist between plant parasitic nematodes and their hosts can be considered to be rather unique, perhaps in the larger sense only just a degree below symbiosis. Nematode infestation seldom results in outright killing of the host with populations as observed under field conditions. A very low initial inoculum level is known to have even stimulatory effects on the host plants.

The degree of parasitism can be considered to be most advanced when the nematodes stimulate changes in the host cells vital to their own growth and development, without the destruction of the infested cells. The citrus nematode, *Tylenchulus semipenetrans* establishes a group of six to ten 'nurse cells' in the cortex rich in metabolites. Similarly the cyst-forming (*Heterodera* spp., *Globodera* spp.) and the root-knot nematodes

(*Meloidogyne* spp.) produce a group of 'giant cells' around their head end in the stelar region of the hosts. In addition, the root-knot nematode induces extensive pericycle hyperplasia and cortical hypertrophy which result in the formation of galls. In contrast to these physiologically specialised nematodes, the activity of less specialised nematodes may affect the host cells drastically. Removal of cell contents following stylet penetration usually results in collapse of tissues and death of cells.

The deleterious effect on hosts due to nematode infestation is, by and large, physiological. Such physiological changes in hosts are reflected in increased respiration, accelerated enzymic activity of oxido-reductase and hydrolases, increased protein, DNA and RNA synthesis and changed rate of photosynthesis. The exact mechanism how plant parasitic nematodes induce physiological imbalance in the host is rather intriguing and remains unanswered. However, evidence is fairly strong to suggest involvement of plant growth regulatory substances as in the case of the root-knot disease. A number of auxins like indole acetic acid, indole butyric acid, indole acetonitrile and indole acetic acid ethylester are reported from different species of *Meloidogyne*. It is possibleth at auxins of nematode origin are directly injected into the host tissue to bring about disturbed metabolism and cellular abnormality. Alternatively, it has been suggested that proteolytic activity from root-knot nematode larvae may break down plant proteins, subsequently leading to the accumulation of IAA, and this may result in the formation of galls. On the other hand, resistance of host plants to nematode invasion is characterised by hypersensitive reactions, failure of the parasite to establish permanent feeding sites (giant cells in *Heterodera, Globodera, Meloidogyne,* and nurse cells in *T. semipenetrans)* and thickening of cell wall around nematodes head. Differences in phenolic substrates in the hosts or in B-glucosidases in the parasites have been accounted for resistance. Fate of nematodes in a resistant plant is reflected in retarded rate of reproduction, delayed maturity of the parasite, production of more males and failure of nematode to complete its development.

Control method : All control methods are directed to bring down the nematode population low enough to raise the crop profitably. The choice of a particular method will primarily depend upon the nematode species and the crops involved. The following is a brief outline of different control methods against plant parasitic nematodes.

Method belonging to culture : The objective of these measures is to keep a field free of susceptible plants or adopt suitable land management practices so that nematodes will be reduced to a very small number. This includes crop rotation, fallowing and ploughing during the summer months, mulching and use of resistant or tolerant varieties.

(a) *Rotation of crop :* This is one of the recognised methods of keeping nematode populations under check, especially where they have distinct host preferences, e.g. in the case of cyst-nematodes of potatoes, wheat, barley, etc. Susceptible crops should be grown only once in a few years, rotating them with highly resistant varieties or non- host crops. Vegetable crops should not be grown repeatedly in the same field, but should be rotated with cereal crops in order to reduce infestation by root-knot nematodes.

(b) *Summer fallow and ploughing :* Hot dry summer conditions prevalent in India could be made use of for nematode control with very satisfactory results. Nematodes are very sensitive to heat and the drying action of the sun and the wind. Keeping the land fallow during summer months and ploughing the soil deep 2-3 times at intervals of 10-15 days is an excellent method to reduce root-knot nematodes and others to the minimum.

(c) *Use of organic amendments* : Soil amendment with green manure, compost, oil-cakes (of *neem mahua,* mustard, groundnut, cotton, linseed, *karanj),* saw dust, etc. has been found to reduce nematode populations. *Neem, karanj* or groundnut cakes incorporated into soil at the rate of 1,500 kg/ha gives good control of root-knot and other plant nematodes and could be practised wherever possible. Apart from encouraging the multiplication of natural enemies like nematode-trapping fungi, the decomposition products of these organic amendments are toxic to nematodes.

(d) *Resistant use and tolerant varieties* : This is one of the most practical methods of controlling nematodes and most economical from the farmers' point of view. However, the rare availability of good sources of resistance, the long period required to evolve a commercially acceptable resistant variety, and the possible evolution of resistance-breaking bio types in some species of nematodes are some of the practical difficulties encountered in adopting this method. Recent experiences emphasise the use of polygenic and combined resistance against nematodes and other pathogenic microorganisms like fungi since both nematodes and fungi are known to be intimately associated in disease complexes. A full and long range collaboration between plant breeders, nematologists and plant pathologists is essential for successful utilization of this method against phytophagous nematodes.

A number of resistant varieties of plants have been evolved by workers in different parts of the world against specific nematodes. These include tobacco varieties 'NC 95', 'Coker 254' and 'Coker 258'; cotton varieties 'Darwinii', 'Auburn 56', 'Dixie King 11'; soyabean varieties 'Habro', 'Jackson', 'Hardee'; alfalfa varieties 'African Moapa', 'India'; cowpea varieties 'Iron',

'Clay', 'Grand'; lima bean variety 'Nemagreen'; capsicum varieties 'Red chilli,' 'Mississippi', 'Nemaheart'; tomato varieties 'Hawaii 5229,' 'Nematex', 'VEN 8' ; sweet potato variety 'Nemagold', etc., against the root-knot nematode *M. incognita.* The release of the resistant tomato variety 'SL 120' by the, Indian Agricultural Research Institute against *M. incognita* deserves special mention here. Likewise, the soyabean variety 'Peking' is reported to be resistant against the cyst nematode. *Heterodera glycines* and the reniform nematode *Rotylenchulus reniformis,* and potato varieties 'Antinema', 'Cobra', 'Ulster Glade' etc., against some biotypes of the golden nematode of potatoes, *Globodera rostochiensis.* Sustained efforts to screen wheat and barley germplasm against the cereal cyst nematode, *Heterodera avenge* is likely to yield fruitful results in the near future. Some of the barley varieties like 'Morocco' and 'Moroccaine' have already shown great promise as sources of resistance against this nematode.

Biological control method : Plant parasitic nematodes have their own natural enemies including predatory nematodes and fungi, protozoa, arthropods and viruses, and these offer as possible biological agents to control obnoxious nematodes. To date, such control measures remain of academic interest and very little progress has been made in this direction.

Physical control method : The most commonly used physical methods of nematode control are steam sterilization of soil and hot water treatment of planting material. Hot water treatment is widely used for transplanted crops. Effective times and temperatures vary from 240 minutes at 43.3°C to 1 minute at 54.4°C. Usually dormant bulb material like narcissus, iris, chrysanthemum, ester lily, and root stocks like citrus, peach etc., are treated against nematode parasites in some of the western countries. Hot water treatment of rice seed at 52°-55°C for 10-15 minutes is a very effective method to control the white tip nematode, *Aphelenchoides besseyi,* which is transmitted through seed. The control of the ear cockle and 'tundu' diseases of wheat by removal of nematode galls from seed material by dry cleaning or floatation techniques is another example where physical methods have given excellent results all over the world.

Chemical control method : This is by far the most effective method to control plant parasitic nematodes. The most successful nematicides are soil fumigants. They may be applied directly as liquids, emulsions or granules, but the active ingredient spreads in the soil in the form of a gas. The halogenated hydrocarbons D-D (a mixture of 1,3-dichloropropene and 1,2-dichloropropane), EDB (1,2-dibromo-ethane), MBr (methyl bromide) and DBCP (1, 2-dibromo-3 chloropropane) are among the most commonly used soil fumigants, marketed under different trade names by manufacturers. Several other allied formulations are also available. For successful results, it is essential that the soil to be treated is well prepared to a seedbed condition, soil moisture is moderate and uniform, and the soil temperature is between 15°-30°C at the time of application.

Soil fumigants are injected into the soil at a depth of 20-25 cm and at a spacing of about 30 cm. Applicators of various types are in use in many countries, ranging from hand-injectors suitable for home gardens to large tractor-drawn multi-chisel machines capable of covering 2-4 hectares a day. Soil fumigants like D-D; EDB etc., should be applied at least 3-4 weeks prior to planting as they are highly phytotoxic. However, DBCP can be used in irrigation water around the roots of living plants as they are well tolerated by most crop plants at the dosages generally recommended.

A number of organo-phosphorus derivatives and carbamates also possess nematicidal properties to varying degrees and many of them are systemic in action. Some of them are widely used as insecticides also. Phorate, Fensulfothion, Thionazin, Dichlofenthion, Disulfoton, Diazinon, Fenamiphos etc., among the organophosphates, and Al-dicarb, Carbofuran, Methomyl and Vydate among the carbamates have been widely tested against nematodes all over the world. They are generally characterized by being highly active, low in fumigant activity and by their non-persistence. One of the great advantages of these compounds is that they can be applied as granules in the row or in the planting holes at the time of planting. Bare root dips of transplants and seed treatments are other possibilities still in the experimental stage, though there may be phytotoxicity hazards under certain circumstances. Vydate is perhaps the first chemical which has given good nematode control when used as a foliar spray, but has not become popular owing to its high mammalian toxicity and short duration of the protection achieved.

Detection of nematode problems and adoption of appropriate control methods at the proper time is one of the modern methods of increasing the yields of many of our agricultural and horticultural crops. In general it may be stated that chemical control of nematodes has not made much headway in India due to the prohibitive cost of nematicides and the difficulty in application over large areas. In spite of the high cost, a judicious use of nematicides would be economical in commercial crops like cotton, tobacco, sugarcane, grapevine, citrus, vegetables etc., in heavily infested fields where lack of such control would mean very poor or no crops. In many situations an integrated approach involving more than one method like crop rotation, deep summer ploughing, chemical treatment of nursery beds or bare root dips of transplants, and use of resistant or tolerant varieties offer the best promise.

A list of common nematode problems in our important crops and their control is given below.

7

Plant Physiology

Water Absorption and Movement in Plant

- Water is the main source of plant and is essential for the maintenance of life, growth and development.

Different Processes

Imbibition Process

- This is the process of absorption of water by the particles of solid substances without forming a solution.
- Imbibition helps in the germination of seeds movement of water from cell to cell in the plant body and the ripening of seeds.

Diffusion Process

- This is the movement of molecules of gases, liquids and solutes from the higher concentration to the lower concentration area.
- The rate and direction of diffusion are influenced by the temperature and pressure.

Osmosis Process

- In the osmosis the movement of solvent molecules from a region of its high free energy to a region of its low free energy when the two are separated by a semipermeable membrane till a state of equilibrium is reached.
- Osmosis is of two types-endosmosis and exosmosis.
- Osmosis plays an important role in the absorption of water by the root-hairs, cell to cell movement of water, induction of turgidity, providing mechanical strength, induction of turgidity.
- Osmosis helps the plants in developing resistance to drought injury, movement of plant parts and opening and closing of stomata.

Plasmolysis Process

- This is the contraction and separation of protoplasm from cell walls due to exosmosis.
- The plasmolysis process is utilized in daily life in the preservation of meat, jellies and food stuffs and in preventing the growth of plants in the cracks of walls.
- The total amount of water present in the soil is called holard.
- The water available to the plant is called chesard.
- The water which cannot be absorbed is called echard.
- The symptoms appearing in a plant or the plant part due to the scarcity of water are known as wilting. Wilting may be temporary or permanent.
- Plants absorb water by two distinct mechanisms which independently operate. These mechanisms are active absorption and passive absorption.
- Active absorption of water involves the utilization of metabolic energy. Water is absorbed by the activity of root hairs.

- Passive absorption of water accounts for about 98% of the total water uptake by a plant. This process does not involve the expenditure of metabolic energy.
- Root pressure, guttation and bleeding are example of active absorption in plants.
- Root pressure is the pressure developed in the xylem vessels as a result of metabolic activity of roots.
- Under conditions of water scarcity and low temperature the roots develop negative pressure.
- Guttation is the exudation of water drops from the edges or tips of the leaves.
- It occurs in the plants, growing under conditions favouring high root pressure and low transpiration, through the structure called hydathodes.

Bleeding

- It is the exudation of sap from the injured parts of the plants.
- The upward transport of water from the roots to the aerial parts of plants is called ascent of sap.

Cohesion

- Cohesion tension and transpiration pull theory as proposed by Dixon and Jolly.
- It is widely accepted theory to explain the mechanism of ascent of sap in plants.
- The loss of water in the form of vapours from the living tissues of aerial parts of the plants is called as transpiration.
- Approximately 98% of the water absorbed by the land plants is lost in transpiration.
- Three kinds of transpiration are distinguishable-stomatal, cuticular and lenticular.

- Stomatal transpiration accounts for about 90% of the total water loss from the plant by transpiration.
- The opening and closing of stomata is controled by the size and shape of guard cells resulting from the changes in their turgor.
- Transpiration is influenced by environmental factors such as-light, temperature, atmospheric humidity, wind velocity etc. and internal factor also influenced of transpiration.
- Transpiration is a necessary evil because of its advantages and disadvantages.
- Antitranspirants are the chemical substances that reduce the transpiration rate without effecting the other metabolic activities of the plants.

Process of Photosynthesis

"Photosynthesis is the process by which the green chlorophyll containing cells of the plants manufacture the carbon compounds with the help of inorganic raw material such as CO_2 and water in the presence of sunlight."

- The atmosphere contains only about 0.03% CO_2 by volume.
- The earth surface receives about 650×10^{21} cal of sunlight energy per year.
- It is estimated that only 1.2% of this total energy is utilized in photosynthesis.
- All green parts of the plants contain chloroplasts and leaves are the main sites of photosynthesis.
- Chloroplast pigments are of three types :

Chlorophylls

- This is green photosynthetic pigments.

Carotenoids

- This is yellow, orange or red coloured pigments and absorb primarily the violet to blue regions of the specturn.

Phycobilin

- According to the modern views, photosynthesis is fundamentally an oxidation-reduction process in which H_2O is oxidised to O_2 and CO_2 is reduced to carbohydrates and the radiant energy of the sun is stored in the form of chemical energy.
- According to the modern concept, photosynthesis is completed in two distinct steps namely light reaction and dark reaction.
- In the light reaction, ATP and $NADPH_2$ are generated and the water molecules undergo splitting resulting in the evolution of molecular oxygen.
- Photosystem-I is associated with both cyclic and non-cyclic photophosphorylation and reduction of NADP to $NADPH_2$.
- While photosystem II is concerned with only non-cyclic photophosphorylation and photo-oxidation of water resulting in the evolution of molecular oxygen.
- Dark reaction is also known as black man reaction in honour of the scientist who reported it first.
- The biochemical reaction of the "dark reaction phase" also known as Calvin cycle takes place in the stroma part of the chloroplast where all the necessary enzymes are located.
- Dark reaction primarily consists of three steps :

  (i) carboxylation

  (ii) glycolytic reversal and reduction

  (iii) regeneration of RuBP

- Photorespiration is the process of uptake of O_2 and production of CO_2 in light by photosynthesising tissues.
- Photorespiration is the common in C_3 plants.
- Photorespiration is also called as C_2 cycle because of 2-carbon intermediates.
- C_4 plants (Sugarcane, maize, sorghum etc.) Partially overcome the disadvantage of photorespiration.
- C_4 plants show a characteristic Kranz anatomy of the leaves.
- Parts of the plant such as the green tissues (leaves) which synthesis the carbohydrates are known as "Source".
- The parts of the plants such as non-green tissues where carbohydrates are utilized for maintenance growth and development are called as "Sink".

Bibliography

Agassi, J. : *Animal Husbandry and Rural Development,* Oxford, New York, 2002.

Alfred, D. : *Pest Management,* Himalaya Darshan Prakashan, Rishikesh, 1980.

Anderson, Edgar : *Plants, Man and Life,* University of California Press, California, 1967.

Arshady, R. : *Man's Role in Changing the Face of the Earth,* University of Chicago Press, Chicago, 1965.

Bandyopdhyay : *A Textbook of Agro-forestry with Applications,* Vikas Publishing House, New Delhi, 1992.

Bansil, P.C. : *Agricultural Problems of India,* Vikas Publishing House, New Delhi, 1977.

Bhatia, B.M. : *Poverty, Agriculture and Economic Growth,* Vikas Publishing House, New Delhi, 1977.

Billihgs, W.D. : *Plant, Man and the Ecosystem,* MacMillan, London, 1971.

Bose, A. : *India's Urbanization,* Kanishka Pub., New Delhi, 1980.

Buchanan, K. : *India : The Land and People,* Crown, New York, 1981.

Chintamani Raghuraj : *Advances in Agro-Industry and Food Processing,* Dominant Books, New Delhi, 1999.

Chisholm, Michael : *Rural Settlement and Land Use : An Essay in Location,* Hutchinson, London, 1966.

Coggin Briwn, J. and Dey, A.K. : *India's Mineral Wealth,* Oxford University Press, New York, 1990.

David, G. : *Principles of Crop Production : Theory, Techniques and Technology,* MacMillan, London, 1999.

Dayal, R. : *Precludes Fungi,* Commonwealth Publishers, New Delhi, 2000.

Dhawan, B.D. : *Irrigation in India's Agricultural Development,* Commonwealth Publishers, New Delhi, 1994.

—— : *Studies in Irrigation and Water Management,* Commonwealth Publishers, New Delhi, 1989.

Dikshit, R.D. : *Geography and the Teaching of the Land and Agriculture,* Department of Geography, Poona, University, 1984.

Dubey, R.S. : *Agricultural : Issues and Application,* Gyan Books, New Delhi, 1987.

Durga Prasad, M.K. : *India : Inland Water Resources,* Discovery Publishing House, New Delhi, 1999.

Frankel, F.R. : *India's Green Revolution : Agriculture Costs of Land Growth,* Princeton University Press, Princeton, 1971.

Gautam, N.C. and Narayan, L.R.A. : *Wastelands in India,* Pink, Mathura, 1988.

Hall, A.D. : *An Introduction to the Science of the Nutrition of Plants and Animals,* Discovery Publishing House, New Delhi, 1993.

Hourani, G.F. : *Arab Seafaring in the Indian Ocean and Early Medieval Times,* Princeton University Press, Princeton, 1951.

Hussain, M. : *Systematic Agricultural Geography,* Rawat Publications, Jaipur, 1999.

Isser, G.P. : *Changing Face of Poverty and Globalisation,* Gyan Books, New Delhi, 2002.

Joshi, H.L. : *Agricultural of India,* Rawat Publications, Jaipur, 1990.

Karanth, Apoorva : *Plant Genetics and Food Crop Science,* Dominant Books, New Delhi, 1999.

Kumar, T. : *History of Rice in India : Mythology, Culture and Agricultural,* Gyan Books, New Delhi, 1988.

Kuriyan, G. : *India : A General Survey,* National Book Trust, New Delhi, 1970.

McRae, S.C. and Burnham, C.P. : *Land Evaluation,* Oxford University Press, New York, 1981.

Mishra, S.R. : *Bacterial Plant Diseases,* Discovery Publishing House, New Delhi, 2003.

—— : *Mallicutes and Plants Diseases,* Discovery Publishing House, New Delhi, 2004.

Palanisami, K. : *Tank Irrigation in 21st Century,* Discovery Publishing House, New Delhi, 2000.

Pannikkar, K.M. : *India and the Indian Ocean : An Essay on the Influence of Sea Power on Indian History,* George Allen and Unwin, London, 1951.

Pedelaborde, P. : *The Monsoon,* Methuen and Co., London, 1963.

Rao, K.L. : *India's Water Wealth,* Orient Longman, New Delhi, 1975.

Robert, W. : *Middle America : Its Land and Peoples,* Prentice Hall, Englewood Cliff, 1989.

Satish, S : *People's Participation and Irrigation Management,* Commonwealth Publishers, New Delhi, 1990.

Shandhya, T.K. : *Agricultural Credit and Nabard,* Deep & Deep Publications, New Delhi, 2003.

Sharma, D.N. : *Minerals of India,* National Book Trust, New Delhi, 1994.

Sharma, S. S. : *Sustainable Agriculture : Poverty and Food Security,* Rawat Publications, Jaipur, 2002.

Sharma, V.K. : *Trees and Protection of Environment,* Deep & Deep Publications, New Delhi, 2004.

Singh, D.K. : *Technological Change and Agricultural Development,* Deep & Deep Publications, New Delhi, 1993.

Singh, Harjinder : *Agricultural Problems in India,* Gyan Books, New Delhi, 1988.

Singh, L. : *Green Revolution and Cropping Pattern,* Deep & Deep Publications, New Delhi, 1993.

Somani, K.K. : *Indian Economics and Social Traditions,* Gyan Books, New Delhi, 2002.

Subrahmanya, S. : *Studies in Agricultural Development,* Deep & Deep Publications, New Delhi, 1996.

Sundra, B. : *Sugarcane Cultivation,* Uppal Publishing House, New Delhi, 1999.

Thakur, D. : *Politics of Land Reforms in India,* Commonwealth Publishers, New Delhi, 1989.

——: *Trends in Agro-Industrial Sector,* Deep & Deep Publications, New Delhi, 2002.

Valcarcel, M. : *Weather and Agriculture,* Oxford University Press, London, 1997.

Verma, M. : *The Survey of Indian Agriculture,* Kasturi & Sons, New Delhi, 1999.

Vyas, V.S. : *Policies for Agricultural Development,* Rawat Publications, Jaipur, 1997.

Whalley, P. B. : *Studies in Green Revolution,* United Asia Publications, Bombay, 1970.

Wismer, R. : *Eco-Science,* W.H. Freeman & Co., San Franciscess., 1977.

Zebrowski, E. : *The Arid Lands : Their Use and Abuse,* Longman, New York, 1983.

Zschunke, N. : *Insect Control,* McGraw Hill, New York, 1970.